AF709799

Geographers

Geographers: Biobibliographical Studies

This volume is part of a series of works, published annually, on the history of geography undertaken on behalf of the Commission on the History of Geography of the International Geographical Union and the Commission of the International Union on the Philosophy and History of Science. *Chair*: Marcella Schmidt di Friedberg, University of Milan–Bicocca, Piazza dell'Ateneo Nuovo, I–20126, Milano, Italy. *Other full members*: Marie-Vic Ozouf-Marignier, EHESS 190–198 Avenue de France 75244, Paris cedex 13, France; Federico Ferretti, University College Dublin School of Geography, H015 Newman Building, Belfield–Dublin 4, Ireland; Elizabeth Baigent, School of Geography and the Environment, University of Oxford, South Parks Road, Oxford, OX1 3QY, UK: André Reyes Novaes, Universidade do Estado do Rio de Janeiro, Departamento de Geografía Humana, Rua São Francisco Xavier 524, Maracanã 20550-013–Rio de Janeiro, Brazil; Guy Mercier, Centre interuniversitaire d'études sur les lettres, les arts et les traditions (CELAT), Département de géographie, Université de Laval, Pavillon Charles-De Koninck 19, Local 6259, Québec G1K 7P4, Canada; Fukuda Tamami, School of Environmental System Sciences, Osaka Prefecture University 1-1, Gakuen-cho, Naka-ku Sakai, Osaka 599–8531, Japan; Leon Yacher, Department of Geography, Morrill Hall 118A, Southern Connecticut State University, 501 Crescent Street, New Haven, CT 06515–1355, USA; Perla Zusman, CONICET/ Instituto de Geografía, Universidad de Buenos Aires, Puán 480, 4to piso CP 1406, Ciudad Autónoma de Buenos Aires, Argentina; Bruno Schelhaas, Leibniz-Institut für Länderkunde, Schongauerstraße 904328 Leipzig, Germany; Judite Nascimento, Departamento de Ciència e Tecnologia, Universidade de Cabo Verde, Campus do Palmarejo Praia Santiago, Cape Verde. *Honorary Chairs*: Vincent Berdoulay, Département de Géographie, Laboratoire Société, Environnement, Territoire, Université de Pau et des Pays de l'Adour, Domaine Universitaire, 64000 Pau, France; Jacobo García-Álvarez, Departamento de Humanidades: Historia, Geografía y Arte, Universidad Carlos III de Madrid C/ Madrid 126, Edificio 14, despacho 14.2.13, 28093 Getafe, Spain. Other honorary members: Michael Heffernan, School of Geography, University of Nottingham, University Park, Nottingham NG7 2RD, UK; Jean-Yves Puyo, Département de Géographie Laboratoire Société, Environnement, Territoire, Université de Pau et des Pays de l'Adour, Domaine Universitaire, 64000 Pau, France; Charles W. J. Withers, Institute of Geography University of Edinburgh, Drummond Street, Edinburgh EH8 9XP, Scotland, UK; João Carlos Garcia, Departamento de Geografía, Facultade de Letras, Universidade do Porto, Via Panorâmica s/n, 4150–564 Porto, Portugal; Jan Vandersmissen, Centre d'Histoire des Sciences et des Techniques, Université de Liège 17, place Delcour 4020, Liège, Belgium.

Geographers

Biobibliographical Studies

Volume 38

Edited by
Elizabeth Baigent and André Reyes Novaes

on behalf of the
Commission on the History of Geography
of the International Geographical Union and the
International Union of the History and Philosophy of Science

BLOOMSBURY ACADEMIC
LONDON • NEW YORK • OXFORD • NEW DELHI • SYDNEY

BLOOMSBURY ACADEMIC
Bloomsbury Publishing Plc
50 Bedford Square, London, WC1B 3DP, UK
1385 Broadway, New York, NY 10018, USA

BLOOMSBURY, BLOOMSBURY ACADEMIC and the Diana logo are trademarks of Bloomsbury Publishing Plc

First published in Great Britain 2020

Copyright © Elizabeth Baigent, André Reyes Novaes and Contributors, 2020

Elizabeth Baigent and André Reyes Novaes have asserted their right under the Copyright, Designs and Patents Act, 1988, to be identified as Editors of this work.

All rights reserved. No part of this publication may be reproduced or transmitted in any form or by any means, electronic or mechanical, including photocopying, recording, or any information storage or retrieval system, without prior permission in writing from the publishers.

Bloomsbury Publishing Plc does not have any control over, or responsibility for, any third-party websites referred to or in this book. All internet addresses given in this book were correct at the time of going to press. The author and publisher regret any inconvenience caused if addresses have changed or sites have ceased to exist, but can accept no responsibility for any such changes.

A catalogue record for this book is available from the British Library.

A catalog record for this book is available from the Library of Congress.

ISBN: HB: 978-1-3501-2797-5
ePDF: 978-1-3501-2798-2
eBook: 978-1-3501-2799-9

Series: Geographers: Biobibliographical Studies

Typeset by Deanta Global Publishing Services, Chennai, India
Printed and bound in Great Britain

To find out more about our authors and books visit www.bloomsbury.com and sign up for our newsletters.

Contents

Contributors

Elizabeth Baigent graduated BA, MA and DPhil in geography from the University of Oxford and is Reader in the History of Geography there, having been Research Director of the *Oxford Dictionary of National Biography* from 1994 to 2004.

Hugh Clout is Professor Emeritus in the Department of Geography at University College London, United Kingdom. He is a fellow of the British Academy.

Claire Hann graduated BA in geography from the University of Oxford (1998), and MA and PhD in urban history from the University of Leicester. She is Equality and Diversity Officer and Researcher at the School of Geography and the Environment, University of Oxford.

Rogério Haesbaert is Professor and Chair of Human Geography in the Institute of Geosciences at Universidade Federal Fluminense, Brazil. He has published extensively on political geography and multi-territoriality. He did his postdoctoral research at the Open University supervised by Doreen Massey.

Ana Angelita da Rocha is Senior Lecturer of Geography Education at Federal University of Rio de Janeiro. Her postdoctoral research theme was Doreen Massey's lessons for a political agenda of school Geography, supervised by Dr Rogério Haesbaert.

Diogo Gaspar Silva holds a degree in geography from the Institute of Geography and Spatial Planning at the University of Lisbon and is now studying for his master's degree in human geography there. His work focuses on the trends of social and spatial transformation of urban spaces.

Susan Squibb graduated BA in 1973 from Oxford's School of Geography and is a Chartered Librarian who worked with the school's collections for 25 years.

She graduated MA in English local history from the University of Leicester (1992) and in 2018 began DPhil research in the School of Geography.

Sarah Tyacke is a historian with special interests in the history of cartography and travel. She served for many years in the map room of the British Museum/ British Library and from 1991 to 2005 was Keeper of Public Records and Chief Executive of the Public Record Office of the UK, and in this role oversaw the office's transition to become the new National Archives in 2003.

Heather Viles graduated BA and MA in geography from the University of Cambridge, and DPhil from the University of Oxford. She is Professor of Biogeomorphology and Heritage Conservation at Oxford's School of Geography and the Environment of which she was head 2015–2019. Her research interests focus on the application of science to heritage conservation.

I am not a man: Maybe this makes things different

Doreen Massey, cited in Albet and Benach 2012, 49–50

Elizabeth Baigent and André Reyes Novaes

Thus the highly distinguished geographer Doreen Massey, memorialized in this volume of *Geographers: Biobibliographical Studies* (*GBS*), mused on her life and geographical work. Revealingly, she did not say simply 'I am a woman'; rather, being a woman, she defined herself against the prevailing male norms in her – our – discipline and quotidian life. It is with pleasure that we use her comment as an epigraph to the first volume of *GBS* in which all subjects are female: Doreen Massey, Helen Wallis, Alice Saunier-Seïté, Isabel Margarida André, and Fanny Herbertson, Nora MacMunn, Marjorie Sweeting, Mary Marshall, Barbara Kennedy and other Oxford's women geographers who are memorialized in a group article. As always, however, Massey's thoughts cannot be lazily co-opted in support of one's own endeavours, for she also declared, as her memoir here explains, that only her work, not her memoirs, mattered. We have then our themes for the introduction to this volume: How has geographers' gender shaped their lives and our discipline? Does gender continue to shape professional lives and practice in our discipline, and if so how? Can the biographical memoir – a genre still dominated by the ideal of the Great Man – usefully bring to light other kinds of lives? Can the biographical memoir illuminate the work, or does it, at best, simply clutter our recollection of it with the transient and the trivial and, at worst, detract attention from the work?

To the first topic: how has geographers' gender shaped their lives and our discipline? In many ways, the answer to this question is so clear as not to need restating. Wilbur Zelinsky, then President of the Association of American Geographers (AAG) memorably addressed his association in 1973 on 'The strange case of the missing female geographer' (1973a and b), something that he found utterly reprehensible as well as simply strange. The paucity of women geographers was one problem: others are exemplified by the lives of the women memorialized here. Women geographers were not promoted as rapidly as men: Marjorie Sweeting, who was never made a professor despite a highly distinguished research record, exemplifies this. Women but not men were often expected to choose between a 'normal' domestic life and a career: the childless Wallis, Massey and Sweeting exemplify this, though André proudly bucked the trend and her experience shaped her work. Women geographers are overrepresented in roles often looked down on in the academy: the school teachers, cartographers and librarians included here exemplify this. (Anyone who doubts how inappropriately dismissive academics can be of these roles will note the astonishment of Wallis's university tutor that she preferred a post in the British Museum, with its map collection of very considerable international importance, to one in a geography department fast becoming a byword for parochialism. Of course, the fact that women were overrepresented in roles such as librarianship confirmed the undesirability of those roles in the eyes of many men.) Women and men were appraised according to different criteria, in Saunier-Seïté's case, for example, for her perfume, charm and attractiveness. Women's daily lives seem different from men's: Massey regretted it when a very well-known and influential male academic was cited as a role model as she thought the daily realities of life would make it impossible for most people to emulate him. 'When', she once asked, 'does he do the washing-up?' (cited in Lee 2018, 150).

Gender difference in past geographical lives is obvious, and several of our subjects drew from this experienced difference to reject the alleged gender neutrality of geographical texts. Massey argued forcefully for the gender specificity of many of the processes she observed and relentlessly exposed the gendered nature of analyses, including those that were in other ways fully aligned with her politics. André's biographer shows how her academic analyses of the differential impacts of politics, religion and other phenomena on women's lives were rooted in her experience of gender differences in daily life. It is clear that we still do not know enough about the lives of past women geographers, despite the efforts of those such as Avril Maddrell (2008, 2009) in a British context, Nicolas Ginsburger (2017) in a French context, Janice Monk

(2003, 2004, 2006, 2017) in a US context, and women's biographers in *GBS*. Moreover, it is clear that, as we write the history of women geographers, we thereby expand the history of the 'geographical canon' (Monk 2012; Maddrell 2009, ch. 4, 2015) and rewrite the history of 'geographical work': Maddrell (2008) and Monk (2017), for example, considering past women geographers in military intelligence, government bureaux and libraries, rather than academe.

To our second topic: does gender continue to shape professional lives and practice in our discipline, and if so how? Following Zelinsky's 1973 call for action, the AAG established a Committee on the Status of Women, which undertook systematic survey of US geography departments. Its results were revealing, if not surprising (Golledge and Halperin 1983). That committee is now a standing committee of the AAG with a brief constantly to monitor the status of women (http://www.aag.org/cs/about_aag/governance/committees#Committee_on_the_Status_of_Women_in_Geography) and the AAG in various ways continues to monitor and further the status of women (see, for example, 'Women in geography' in its *International Encyclopedia of Geography* (Peake 2017). The International Geographical Union (IGU)'s Commission on Gender and Geography (*est.* 1988) also helps to track the present state of the discipline. As well as providing a place for like-minded geographers (mainly women) to share ideas and find mutual support (Monk 2008), the commission has issued working papers on 'The status of women geographers and of the teaching of feminist geography in academic institutions in the United Kingdom' (Peake and McDowell 1995) and 'Gender representation in academic geography in Turkey' (Özgüç 2006). Studies done under the auspices of the AAG and the IGU extend those of individual geographers (e.g. Monk (2003, 2004 among many other of her studies) focused mainly on the United States; Garcia Ramon, Castañer and Centelles (1988) for Spain; Murata (2005) for Japan; Mackenzie (1989) for Canada; McDowell (1979) for Britain). However, the picture is still sadly limited, as well as being geographically uneven: the Anglophone world is unusually well documented as well as being more privileged in other ways (Peake 2017).

We do have societies for women geographers, of which a remarkable early example is the Society of Women Geographers, founded in 1925 in New York by explorers Marguerite Harrison, Blair Niles, Gertrude Shelby and Gertrude Emerson Sen to encourage women's participation in geography, broadly interpreted to include exploration, fieldwork and geographical investigation (http://www.iswg.org/about/history). Its aims of supporting women practitioners, drawing attention to their achievements and helping them overcome barriers (see Monk 2017 for use of their oral histories) overlap to some extent with those of groups that formed within broader professional geography societies: the Royal Geographical Society with the Institute of British Geographers (RGS–IBG)'s Women and Geography Study Group (*est.* 1980) is the relevant example from the UK. The trend recently has been to move away from a focus on women in geography per se: the RGS–IBG group is now the Gender and Feminist Geographies Research Group http://www.gfgrg.org/about/), while the Arbeitskreis Feministische Geographie (feminist geography study group, *est.* 1989) of the Deutsche Gesellschaft für Geographie (German Geography Society) is now the Arbeitskreis Geographie und Geschlecht (geography and gender study group) (http://ak-geographie-geschlecht.org/). From the Global South we might point to the very active Latin American network of Geography, Gender and Sexuality studies (https://sites.google.com/site/reggsal/; https://reggsila.wordpress.com/). Its journal, *Revista Latino-Americana de Geografia e Gênero*, (*Latin American Journal of Geography and Gender*) speaks of 'focusing specifically on theoretical and methodological perspectives' (http://www.revistas2.uepg.br/index.php/rlagg), that is, it is less concerned with empirical studies of geographical women in the past or present.

As 'the geography of gender' replaces 'women in geography' in some organizations and in many geography courses, there is perhaps a sense that making women visible was yesterday's project. Karen Falconer Al-Hindi applauds the approach taken by contributors to a (very welcome) view of women geographers at the turn of the millennium saying, 'This [2001] effort represents a genuine break with earlier work on the "status of women in geography" (which was mainly empirical) because of its theoretical orientation' (2001, 699). While not disparaging theoretical insights, we see no evidence that conceptually simple but systematic and wide-ranging empirical documentation of the status of women in geography is anywhere near complete (or could be, since constant review is needed). Indeed, more and more hard data are needed to examine assumptions that things will get better (see Lee (1990) for instances when 'things remain the same' and McDowell and Peake (1990) for the 'same old story' for cases when they have not got better); to scotch contemporary suggestions that things have swung so far in women's favour that women are now advantaged rather than disadvantaged in geography as in other public arenas (Falconer Al-Hindi 2001, 697, 701 fn.1); and to raise the alarm when women geographers are not so much 'absent' as 'disappearing' as is reported for Spain (Garcia Ramon and Pujol 2004; Pujol i Estragués 2004).

The lack of systematic data gathering makes policy recommendations harder (cf. e.g. Zelinsky, Monk and Hanson 1982 for advocacy of change on the basis of data) and puts geography at a disadvantage compared with some other disciplines despite the suggestion that women in geography might fare worse than colleagues in other disciplines (e.g. Rubin 1979). The Royal Historical Society, the learned society for professional historians active in or working on Britain, for example, has undertaken two major surveys (2015 and 2018) of the effects of gender on the professional lives of historians in the UK (https://royalhistsoc.org/rhs-report-gender-equality-historians-higher-education/; https://royalhistsoc.org/genderreport2018/) on the basis of which profound differences between men's and women's lives were uncovered and considered recommendations were made for action by departments/faculties, journal editors, conference organizers, teachers/supervisors and others to address the challenges. It would be good if geographical professionals had comparable hard data to reveal their discipline's present and to make conscious attempts to shape its future: exemplary methodological and theoretical perspectives alone will not bring change in our discipline. The need meticulously to document the roles women claimed for themselves and to compare them with those they were 'supposed' to do never ends (O'Hanlon, cited in Levine 2004, 1–2).

To the third topic: can the biographical memoir – a genre still dominated by the ideal of the Great Man – usefully bring to light other kinds of lives? The nineteenth-century biographical memoir was unquestionably shaped by Carlylean notions of the Great Man, but even when that mode was most dominant there was still scope for memorializing women. Often, however, such memorialization reflected the prevailing essentialist claims of the determining effect of sex. Thus while Sidney Lee in *Principles of Biography* (1911, 25–6) declared, 'The aim of biography [of great men] is not the moral edification which may flow from the survey of either vice or virtue; it is the truthful transmission of personality,' Elizabeth Sandford, or as she was styled on the title page 'Mrs

John Sandford', author of *English Female Worthies* (1883), declared, 'To trace the development of character is the chief end in all biography' (cited in Lee 2009, 65). While Sandford was thus forward-looking in claiming that women's lives were worth memorializing despite having been lived largely out of the public domain, her claim rested on the idea that women's sex determined for them different life paths and a different model of biography with a different aim – precisely the moral edification that Lee spurned for men's biographies. Similarly the relationship between the subject and the biographer has been held to be different when both are women: reflecting on her biography of Charlotte Brontë (1857), fellow novelist Elizabeth Gaskell spoke of the need for a 'tender tie' between biographer and subject (cited in Lee 2009, 59). The existence of such a 'tender tie' can be both the casual and essential assumption of a gendered profession and simultaneously a reality in a world where professional circles are still dominated by men (in the case of Gaskell and Brontë, they were, in fact, friends and both had to justify their role as authors, having resorted to publishing under male pseudonyms). Here we make no essentialist claims that biography of or by women is necessarily different from that of or by men, though it may be. Our women's biographies have no didactic purpose (illustrating 'character' for example) peculiar to them; they record women's public lives in the same realms as men's – strikingly so, in fact, as numerous of our subjects were active not only in geography, but in traditional public roles of politics and public policy (Saunier-Seïté), wartime intelligence (Marshall) and covert operations (André); they are inspired by female role models (Massey's teacher 'Miss Leigh') but also male (André's grandfather); they attend and derive very considerable support from exclusively female schools and colleges (St Hugh's College, Oxford, for example, the women's college of Massey, Wallis, Sweeting, Kennedy and authors Baigent and Squibb), but they campaign for them to be open to men (Sweeting) or leave them for avowedly open institutions (Massey); though our subjects are all women, our biographers are women and men, and some subjects were never personally known to the authors, so no 'tender tie' could exist between them.

However, as the lives of Massey, André, MacMunn and Saunier-Seïté show, geography as a practice cannot escape politics. Massey, MacMunn and André were political activists, Saunier-Seïté shaped and implemented educational policy and some of the money which Nora MacMunn earned by teaching geography to men who had Oxford degrees denied to her because of her sex went to the women's suffrage movement. The practice of writing geographical biographies is as political as that of writing other geographical works, and at least some of the authors and editors have an overarching desire to record women's lives, as well as a commitment to record the particular lives here. Collectively and in some instances individually the memoirs here can be read as a manifesto, as well as being inherently of interest. Doreen Massey refused to admit the legitimacy or even the possibility of 'publications and their conclusions' being gender neutral and we follow her lead here. We resist the trend of many biographies to focus relentlessly on women who are of interest to men (Howarth 1993; Lee 2009, 119) and focus on memorializing a broad spectrum of women, including those who contributed behind the scenes (Buxton) and those who did not live up to early promise (Marshall, Doveton). We resist totalizing and universalizing interpretations that treat women as if all were similar, and all 'naturally' had 'feminine qualities' such as empathy, or were uniformly progressive because they were countercultural in pursuing an independent career (Grimshaw 1996, 53; Levine 2004 2–3). Thus we include right-wing Saunier-Seïté and Fascist sympathiser Marie Bentivoglio, as well as progressive André, Massey and MacMunn; those who were widely liked and admired and those who weren't.

An unusually high proportion of our authors in this volume is female (see the data analysis in Baigent and Reyes Novaes 2017 for the general pattern in *GBS*), perhaps a reflection that the project of excavating women's lives, as a political as well as an academic project, matters more to women. The memoir whose form was varied to suit the lives and material of its subjects (the group article on Oxford's women geographers) was written by a team of women who, like their subjects, are drawn from a variety of geographical careers. The group biography or prosopography has in recent decades been loosened from its earlier ties to family biographies, and used to memorialize other more varied groups, in an effort, Lee suggests, to refashion the genre for a democratic age, to uncover pasts that might be otherwise missed, and to challenge the dominant discourse (2009, 126).The group format has previously been used in *GBS* to memorialize men whose contribution as a group was greater than the sum of their individual contributions: Hugh Clout's articles on geographers at Clermont-Ferrand and Vidalian geographers (see *GBS* 35 and 36) come to mind, but their use here was a slight hybrid, given that some of the men were already included in *GBS* as individual subjects. The most innovative use of the model thus far in *GBS* has been in the article on the Hull House geographers by Christine Dando (see *GBS* 33), which combines some narrative history of Hull House with biographies of its female founders and accounts of the cartography produced from there by women, about women's lives. The group article on Oxford geographers in this volume continues in this mould, combining straight full biographies with some sketches of obscure or representative lives – the latter being a particular methodology in collective biography discussed more fully in the Oxford geographers article. Patricia Grimshaw notes that 'in recovering ways to speak interestingly about women in the past, [women's historians have led] important contestations of traditional master narratives' (1996, 35 and see O'Hanlon 1999 for another illustration of this). The combination of the pragmatic search for new methods that fulfil immediate aims but lead consequently to disruption of the canon is the guiding spirit in this volume, and the aim must be to be transformative, not merely additive (Levine 2004). This is shown most clearly in the Oxford women article, which explicitly challenges the canonical story of geography at Oxford and, because that school was the first in the country, in the UK.

The topic of pushing the mode of biography to fit it for the telling of women's lives and the challenging of the canon brings us back to one of the foremost writers of and on biography: Virginia Woolf who most effectively challenged the prevailing Victorian constraints on the subjects and modes of biography (epitomized not least in the work of her father, Sir Leslie Stephen, man of letters, biographer and the founding editor of the *Dictionary of National Biography*). In her essay 'The art of biography' (1942) Woolf writes,

> Biography will enlarge its scope by hanging up looking glasses at odd corners. And yet from all this diversity it will bring out, not a riot of confusion, but a richer unity. And again, since so much is known that used to be unknown, the question now inevitably asks itself, whether the lives of great men only should be recorded. Is not anyone who has lived a life, and left a record of that life, worthy of biography — the failures as well as the successes, the humble as well as the illustrious? And what is greatness? And what smallness? We must revise our standards of merit and set up new heroes for our admiration. (The Death of the Moth, and other essays, by Virginia Woolf, ttps://ebooks.adelaide.edu.au/w/woolf/virginia/w91d/contents.html)

Woolf's reinvention of life writing – her 'new biography' (1927), has limited bearing on *GBS* whose lives are historical rather than literary, but we can learn much from Woolf's insistence that 'small' lives are worth telling, and that hanging up looking glasses in odd corners can be revealing: the article on Oxford's women geographers tells many 'small lives' and certainly shows plenty of odd corners, but our understanding of this, the first geography school in the UK, is the richer for it. Equally importantly, Woolf's practice enlarges the biographical method in a way illuminating for historical as well as literary life writers. Her innovative *Jacob's Room* is a collage of fragments, or 'hints' as she called them (Woolf 1992, 212; Lee 2005). Though a work of literature, it suggests a method for the historical biographer whose project is historical, but who must construct a life out of fragmentary sources and hints – the constant task of those memorializing small and marginal lives. Hermione Lee speaks of the feminist interest in hidden histories, and describes the disinterment of obscure lives and the claiming of new status and significance for women's stories (2009, 127). This volume does both: it disinters various Oxford geographers who are obscure but interesting, but also claims new status for the stories of those like Helen Wallis, who shaped her country's main cartographical collection and the way in which it was studied, or Fanny Herbertson, a prominent geographical textbook author, who was officially never more than the wife of the head of the Oxford school.

And finally to the topic: can the biographical memoir illuminate the work, or does it, at best, simply clutter our recollection of it with the transient and the trivial and, at worst, detract attention from the work? This is a large topic not confined to biographies of women, but needs comment here following Massey's statement that her work and not her life are what mattered. She is effectively paraphrasing Barthes's famous dictum that 'the author is dead', that is, that author and text exist separately and that the former's life cannot legitimately be invoked as commentary on or explanation of the text. The absurdity to which some have taken the 'life explaining the work' trope is wittily illustrated in Nicola Watson's *The Literary Tourist: Readers and Places in Romantic and Victorian Britain* (2006) where geography mystically enables the literary pilgrim to commune with beloved authors and their works at sites associated with them. The practice can invite ridicule, but need not, however. Massey, who declined to write her memoirs, nonetheless gave an unusually large number of interviews about her life and its connection with her work, and speculated on how other people reconciled the two: it was perhaps the vanity of autobiography rather than the value of biography that she rejected (see e.g. Maddrell 2009; Moss 2001, Jones 2018; Millward 2001; Buttimer 1983; and Berman 1984 for general discussions of autobiography in geography). When sensitively and thoughtfully done, the history and geography of a life can be effective organizing concepts for one view of the subject's work and need not stake an indefensible claim to be the only interpretation. In this volume the author of Isabel André's memoir intelligently uses the device of 'personal geography', that is, the people with whom she collaborated, the problems to which she turned her attention, the case studies that sprang to mind as she sought to convey her ideas, to illuminate her professional geography, The author and editors do not claim that André's personal geography explains her work in any facile sense, nor that other organizing concepts of the biography are impossible, nor that the importance of André's work is confined to her life. Rather, they simply recognize that life and the creation of work, if not its meaning, may be reciprocally contingent. What is important here, however, is that life does not asymmetrically account for women's work while men's work somehow floats free of practical constraints. Thus while we here show marriage and parenthood to have been turning points in Fanny Herbertson's life, we also note what has been missed in accounts of her husband's life, that is, that the same marriage and parenthood bring the same compromise (staying in one place and turning to educational writing and training school teachers) in his professional geographical life – though perhaps they do not for the man who does not do the washing-up.

We hope that you will find this collection of essays interesting and will feel that they collectively enrich our understanding of the profession, of geographical enquiry and of biography in geography.

References

Albet, A. and Benach, N. (2012), *Doreen Massey: Un sentido global del lugar*. Barcelona: Icaria, Colección Espacios Críticos.

Baigent, E. and Reyes Novaes, A. (2017), 'Introduction: Retrospect and prospect of our discipline's biographical record', *Geographers: Biobibliographical Studies* 36, 1–10.

Berman, M. (1984), 'On being a woman in American geography: A personal perspective', *Antipode* 6, 61–6.

Buttimer, A. (1983), *The Practice of Geography*. London: Longman.

[Die Deutsche Gesellschaft für Geographie] http://ak-geographie-geschlecht.org/ (consulted 19 February 2019).

Gaskell, E. C. (1857), *The Life of Charlotte Brontë*. London: Smith, Elder & Co.

[Gender and Feminist Geographies Research Group] http://www.gfgrg.org/about/ (consulted 19 February 2019).

Ginsburger, N. (2017), 'Portrait en groupe de femmes-géographes. La féminisation du champ disciplinaire au milieu du xxe siècle, entre effets de contexte et de structure (1938–1960)', *Annales de Géographie* 713, 107–33.

Golledge, R. G. and Halperin, W. C. (1983), 'On the status of women in geography', *Professional Geographer* 35, 214–18.

Grimshaw, P. (1996), 'Female lives and the tradition of nation making', in I. McCalman with J. Parvi and M. Cook (eds), *National Biographies and National Identity: A Critical Approach to Theory and Editorial Practice*. Canberra, Humanities Research Centre, Australian National University, 35–53.

Hindi, K. Falconer Al- (2001), 'Women in geography in the twenty-first century: Introductory remarks: Structure, agency, and women geographers in academia at the end of the long twentieth century', *Professional Geographer* 52, 697–702.

Howarth, J. (1993), 'Associate Editor typescript report', *Dictionary of National Biography Archives*. Oxford: Oxford University Press.

Jones. M. (2018), 'Anne Buttimer's *The Practice of Geography*: Approaching the history of geography through autobiography', *Geografiska Annaler*: Series B, Human Geography, 100, 396–405.

Lee, D. R. (1990), 'The status of women in geography: Things change, things remain the same', *Professional Geographer* 42, 202–11.

Lee, S. (1911), *Principles of Biography*. Cambridge: Cambridge University Press.

Lee, H. (2005), *Virginia Woolf's Nose: Essays on Biography*. Princeton: Princeton University Press.

Lee, H. (2009), *Biography: A Very Short Introduction*. Oxford: Oxford University Press.

Lee, R. (2018), 'Doreen Barbara Massey 3 January 1944–11 March 2016', *Biographical Memoirs of Fellows of the British Academy* 17, 145–78.

Levine, P. (2004), 'Introduction: Why gender and empire?' in P. Levine (ed.), *Gender and Empire*. Oxford History of the British Empire Companion Series, series editor W. R. Louis. Oxford: Oxford University Press, 1–13.

Mackenzie, S. (1989), 'The status of women in Canadian geography', *Operational Geographer* 7, 2–8.

Maddrell, A. (2008), 'The "Map Girls": British women geographers' war work, shifting gender boundaries and reflections on the history of geography', *Transactions of the Institute of British Geographers* 33, 127–48.
Maddrell, A. (2009), *Complex Locations: Women's Geographical Work in the UK, 1850–1970*. Oxford: Wiley–Blackwell.
Maddrell, A. (2015), 'To read or not to read? The politics of overlooking gender in the geographical canon', *Journal of Historical Geography* 49, 31–8.
McDowell, L. (1979), 'Women in British geography', *Area* 11, 151–4.
McDowell, L. and Peake, L. (1990), 'Women in British geography revisited: Or the same old story', *Journal of Geography in Higher Education* 14, 19–30.
Millward, L. (2001), 'Review of P. Moss, ed. *Placing Autobiography in Geography*', *Biography* 24, 935–8.
Monk, J. (2003), 'Women's worlds at the American Geographical Society', *Geographical Review* 93, 237–57.
Monk, J. (2004), 'Women, gender, and the histories of American geography', *Annals of the Association of American Geographers* 94, 1–22.
Monk, J. (2006), 'Changing expectations and institutions: American women geographers in the 1970s', *Geographical Review* 96, 259–77.
Monk, J. (2008), 'Connecting people, places, and ideas: Reflections on the history of the International Geographical Union Commission on Gender and Geography', http://igugender.socsci.uva.nl/documents/IGUGenderCommissionHistoryJune2008.pdf (consulted 19 February 2019).
Monk, J. (2012), 'Canons, classics, and inclusion in the histories of geography', *Dialogues in Human Geography* 2, 328–31.
Monk, J. (2017), 'Washington women: Practicing geography in the US government, 1915–1970s', *Professional Geographer* 69, 683–93.
Moss, P. ed. (2001), *Placing Autobiography in Geography*. Syracuse: Syracuse University Press, 2001.
Murata, Y. (2005), 'Gender equality and progress of gender studies in Japanese geography: A critical overview', *Progress in Human Geography* 29, 260–75.
O'Hanlon, R. (1999), 'Gender in the British empire', in J. M. Brown and W. R. Louis (eds), *Oxford History of the British Empire* vol. 4: *The Twentieth Century*. Series editor W. R. Louis. Oxford: Oxford University Press.
Özgüç, N. (2006), 'Gender representation in academic geography in Turkey', *IGU Commission Gender and Geography Working Papers* 36.
Peake, L. J. (2017), 'Women in geography', in D. Richardson (ed.), *The International Encyclopedia of Geography: People, the Earth, Environment and Technology*. Wiley-Blackwell for the American Association of Geographers online edn, https://on

linelibrary.wiley.com/doi/full/10.1002/9781118786352.wbieg1173 (consulted 21 February 2018).

Peake, L. and McDowell, L. (1995), 'The status of women geographers and of the teaching of feminist geography in academic institutions in the United Kingdom', *IGU Commission Gender and Geography Working Papers* 33.

Pujol i Estragués, H. (2004), 'De la geògrafa absent a la geògrafa desapareguda? El professorat de geografia de les universitats públiques de Catalunya', *Documents d'Anàlisi Geogràfica* 43, 135–52.

Ramon, M. D. Garcia and Pujol i Estragués, H. (2004), 'Women in academic geography in Spain: Towards a masculinization of the discipline?' *Journal of Geography in Higher Education* 28, 111–19.

Ramon, M. D. Garcia, Castañer, M. and Centelles, N. (1988), 'Women and geography in Spanish universities', *Professional Geographer* 40, 307–156.

[*Revista Latino-Americana de Geografia e Gênero*] http://www.revistas2.uepg.br/index.php/rlagg (consulted 20 February 2019).

[Royal Historical Society] (2015), https://royalhistsoc.org/rhs-report-gender-equality-historians-higher-education/ (consulted February 2019).

[Royal Historical Society] (2018), https://royalhistsoc.org/genderreport2018/ (consulted February 2019).

Rubin, B. (1979), '"Women in geography" revisited: Present status, new options', *Professional Geographer* 31, 125–34.

Sandford, E. (1883), *The Lives of English Female Worthies*. London: Longman.

[Society of Women Geographers] http://www.iswg.org/about/history (consulted 19 February 2019).

Watson, N. J. (2006), *The Literary Tourist: Readers and Places in Romantic and Victorian Britain*. Basingstoke: Palgrave Macmillan.

Woolf, V. (1927), 'The new biography', reprinted in A. McNeillie (ed.), *The Essays of Virginia Woolf*, vol. 4. London: Hogarth Press, 1994, 473–80.

Woolf, V. (1942), 'The art of biography', https://ebooks.adelaide.edu.au/w/woolf/virginia/w91d/chapter23.html (consulted 21 February 2019).

Woolf, V. (1992), *Jacob's Room*, ed. Kate Flint. Oxford: Oxford University Press.

Zelinsky, W. (1973a), 'The strange case of the missing female geographer', *Professional Geographer* 25, 101–5.

Zelinsky, W. (1973b), 'Women in geography: A brief factual account', *Professional Geographer* 25, 151–65.

Zelinsky, W., Monk, J. and Hanson, S. (1982), 'Women and geography: A review and prospectus', *Progress in Human Geography* 3, 317–66.

1

Doreen Massey, 1944–2016

Rogério Haesbaert and Ana Angelita Rocha

Figure 1.1 Doreen Massey in London (2004). Photograph by Rogério Haesbaert.

> *For 'a global sense of place' means that any nation, region, city … as well as being internally multiple, is also a product of relations which spread out way beyond it. (Massey 2004: 6)*

Doreen Barbara Massey was an outstanding human geographer who won admirers for her political activism, feminism, and artistic and academic work in the social sciences. She was an anti-essentialist, internationalist and one of geography's foremost public intellectuals who used different media to claim, charismatically, that geography matters when considering alternatives to neoliberal hegemony (Painter 2016, 40). Massey was always concerned with thinking about space as a strategy for formulating alternatives to political practices and actions. Her premature and untimely death caused shock in the geography community and in leftist activists in different parts of the world.

Massey always declined to write her memoirs, considering that only her work was relevant. We nonetheless hope that this biobibliographical memoir helps readers to see her legacy rooted in its social, academic and political contexts. Declaring geography and politics, and space and power inseparable, and contesting frontiers, regional integrations and globalities, Massey considered that 'all of these processes called "global" are founded in specific places and are made global because they are connected through global relations of power' (Albet and Benach 2012, 8).

For this reason, Massey advocated considering the concrete politics of place, to show confrontations on multiple scales and dimensions, including struggles of identity, because it is impossible to consider, for example, the spatial division of work without discussing the issue of women, or to separate the dilemma of British miners (whose prolonged and bitter strike she strongly supported in 1984–85) from the places where they lived. Jamie Peck, Marion Werner, Rebecca Lave and Brett Christophers argue:

> Massey was on the frontline in some of the signature struggles against Thatcherism, including the miners' strike of 1984–85 and the municipal–socialist project of the Greater London Enterprise Board, events that in retrospect marked an historical inflection point between a regionalist model of labour organization and the ascendancy of the 'new urban left'. (Peck et al. [eds] 2018, 38)

She rejected a Massey 'school', being concerned only with academic rigour and the geography that comes from political action and the streets. One of her main tools was what she called 'academic activism' (Albet and Benach 2012, 39). Massey dedicated decades to geographical thought (with 27 years at the UK's Open University) as well as to political action in various contexts and various parts of the world, especially in Latin America, which she visited on several occasions. She spoke Spanish fluently and not only lectured at several universities but also supported social organizations. Her work was translated into different languages and published in a variety of media.

Massey's enthusiasm, charisma, energy and sense of humour were communicated in person and in her writing, as two of the most personal of her many obituaries recall: Gillian Rose ('I can hear her voice now, and her laughter' (Rose 2016), and Hilary Wainwright ('How we will miss that chuckle' (Wainwright 2016)). Ironic, a 'stories so-far' storyteller, an interlocutor for artists, a radical provocateur, Massey engaged her readers and her audience through her rich geographical imagination.

Relational thinking has become a sine qua non of interpretations on space. Massey's work *For Space* (Massey 2005), probably the best summary of her intellectual maturity, emphasized this thinking, based on her many experiences, in a space, as she said, always open to building new trajectories. Her physical difficulties probably reinforced her emphasis on the intermingling of trajectories and on indispensable openness in the composition of geographical space. Perhaps also because of this, space was treated as a challenge and she felt strongly that *Geography Matters* (1984).

Early inspirations

> I was born in the north of England, in a working class family, in a large area of state housing projects south of Manchester [Massey was born 3 January 1944 at Wythenshaw, Manchester, to Jack and Nancy Massey]. My mother was an administrative secretary in a factory. She left school at 13; my father left school at 11 and went to work as a groundsman. I say this because I was the first member of the family who went to university. I have a sister who is now a good friend, we travel together, and she also got a degree in geography, in Southampton, but doesn't work with it, though she has the imagination and spirit of a geographer. ... So it is a small family in two senses: we are small in size and only four people – and an enormous yellow cat, who is the boss of the house! I studied in a State school and then I got a scholarship to go to a different school. In 1963, I went to Oxford University, also with a scholarship, which makes it possible for the working class to go to university. I studied Geography. Why Geography? I like Geography because of the range that it studies, the nature and the humanities. Also, because I had in school a teacher who was magnificent, Miss Leigh. A teacher is very

> important because she can change an entire life. Miss Leigh was important to me and to my sister too. (Massey in Haesbaert et al. 2006)

David Featherstone, her former student, recalled Massey's humble origins and her parents' struggles to enable her and her sister to have a better life. Her sister, Hilary Corton, emphasizes that their mother, Nancy, was very keen that her daughters receive a good education. Their mother had had to leave school when her father died quite young and she became the family breadwinner. Their father, Jack, also left school early but both parents continued their education at night school while they worked. Massey studied at Sharston Primary School and at Manchester High School for Girls, a well-known and highly academic school.

From an early age Massey had a passion for geography, praising the role of good teachers and education. She was a child fascinated by maps and globes (a little metal globe stood by her desk). 'I know', said Massey, 'but I cannot explain why, that as a child I was incredibly curious about other places. I was lucky enough to live in a household where there was a globe and there was an atlas and I used to spend hours gripped by playing with these things and reading them' (Bond and Featherstone 2009, 403). According to her sister Hilary, the family also had a Chambers English Dictionary and works from the literary canon: the Brontë sisters, George Eliot and Robert Louis Stevenson, for example. Massey's favourite book was *Little Women* by abolitionist and feminist Louisa M. Alcott. Massey loved birdwatching and the night sky. We can remember a unique moment at the Lake District admiring for a long time the night sky cut across by the swirl of stars from the Milky Way, searching also for the trajectory of satellites. In a way, we thought, those satellites crossing the sky were also part of multiple networks, on multiple scales, which composed our local place (Rogério, personal knowledge). As for geography, she was always curious about the places she visited.

> Lacking a car, the family travelled 'into town' on the top of a double-decker bus. There were 2 routes from Wythenshawe (the Corporation housing estate where we lived) into Manchester city centre which gave us the chance to see the varying neighbourhoods, from leafy prosperity to the inner-city housing, the playing fields and grand university buildings. Our annual holiday would involve a day's coach ride or train journey to north Wales or Devon. Using public transport was significant and also gave the opportunity to talk to other travellers. (Corton, pers comm.)

Massey claimed in some interviews that having been born in the north of England, into a working-class family without material privileges, made her feel and think about social and geographic inequalities. '[H]ad there not been a welfare state and the hospitals', Massey reflected, 'I would probably not have survived so well. I really feel in a kind of physical, personal way the need for a welfare state, not as a "safety net", but just for ordinary people simply to provide a decent life' (Freytag and Holyer 1999, 85). Although she left Manchester aged 19 when she went up to St Hugh's College, a women's college of University of Oxford, to read geography as the Irene Shrigley scholar, she always considered herself a northerner, and considered that relevant in her choice of career.

> Another interesting fact for my choosing Geography is because I am from an industrial region, particularly textiles, and very young, I experienced inter-regional differences. In Manchester, we have the habit of saying: what Manchester thinks today, London will think tomorrow. Maybe this is a way of compensating for the poverty of Manchester in relation to London. So, regional inequality was experienced still very young, it is part of my life. (Massey in Haesbaert et al. 2006)

Regional inequality in England became a dominant characteristic of her work during its initial phase (1970–1980) and she returned to it repeatedly, as in her co-authored books *Rethinking the Region* (Massey et al. 1998) and *Decentering the Nation: A Radical Approach to Regional Inequality* (Massey 2003a).

Her sister Hilary thought Doreen a born scholar. From the start she loved reading and learning and had a small green child's desk in her bedroom where she could sit and do her school homework (Corton, pers. comm. 2018). One of her pastimes was to spin the terrestrial globe, stop it at some point and imagine what would be happening there at that moment. From that point started the first inspirations for her rich geographical imaginations and her perception of globalization intimately associated to the specific characteristics of places.

Massey had the ability to extract an entire theoretical reflection based on simple, trivial facts, such as the moment that her mother abandoned an old cake recipe, which she and her sister loved, and baked a new cake which they disliked. From that came a discussion on time–space, which cannot be restored and which we cannot impose on others. In *Travelling Imaginations*, from her book *For Space*, her regular trips to Milton Keynes (the site of the Open University where she taught) are the motivation behind 'thinking space

Figure 1.2 Doreen (right) and her sister Hilary (left) on holiday in the Lake District (2010). Photograph by Rogério Haesbaert.

as the sphere of a multiplicity of trajectories, imagining a train journey...as a speeding across on-going stories' (Massey 2005, 119).

Another space that inspired Massey was the Lake District, a heterotopic English landscape, to use Foucault's term, being an unusual region of mountains and lakes in the northwest of the country, which has occupied a prominent place in the English imagination since the Romantic period. Massey used to spend her holidays there (see Figure 1.2) and, according to Hilary:

> Visits to the Lake District were about walking but more importantly, they were a form of retreat from the demands of her world. Sitting on the fellside in Langdale, or on the shores of Morecambe Bay, our conversation could wander, or we could simply look and listen. If there was a football match on [Massey was a loyal fan of Liverpool Football Club], we would be glued to a tiny radio trying to hear the commentary, followed by our 'singing' of 'You'll Never Walk Alone' to the curious but startled sheep gathered about us. But when we got back home, Doreen would disappear for an hour to write up her 'notes'. (Corton, pers.comm. 2018)

Severe bone fragility, the result of a calcium deficiency, meant that from childhood Massey suffered frequent falls and fractures and had to undergo arduous sessions of physiotherapy to build up her strength and mobility. She spent much time in hospital or attending medical appointments and this fact led her to recognize the importance of, and defend, the UK's National Health Service, which since 1947 has provided free health care to those who need it. Having been told that she would be in a wheelchair by the time she was 40, she was determined to remain physically independent and mobile for as long as possible (Corton, pers.comm. 2018). In consequence, she was very disciplined about doing her daily exercises and going walking. As she said: 'Together, my body is a great disaster .... I cannot be exclusively cerebral ... because this body is here every day and, so, it is a very important aspect to me' (Albet and Benach 2012, 62). The result was a debate on the inseparable relation between discursive and non-discursive, the physical and the social.

Refusing to let bodily challenges constrain her, Massey claimed that she was 'one of those geographers who put on boots and go to the country to walk' (Albet and Benach 2012, 60). Despite her difficulties (such as when she broke her foot during her stay in Nicaragua), she enjoyed travelling. She liked to remember a boat trip to the Arctic and a train trip on the Trans-Siberian Railway, which she took with her sister. At heart, she said: 'It is about loving this planet, about being constantly surprised (almost naively) about how incredible it can be.' Despite 'not being religious at all', she claimed that 'there is something pantheist in this appreciation, but it is the world we live in that is delicious, marvelous to me' (Albet and Benach 2012, 60).

Political–academic trajectories

In a conversation about her main inspirations and influences, Massey answered:

> Well they aren't intellectual, and that is what is so interesting ... I have been asked this question before and I couldn't say that there have been particular people or authors that have been guiding lights in that sense. It has been more, and this may be in part a product of the generation I have been part of, that the stimulus to stuff, the reason for asking questions and the ways in which debates got framed, have come out of being part of political movements. Whether that has been in the late 60s and the 70s with the emergence of Marxism, feminism, sexual liberation or the kind of stuff that has happened more recently, and more generally an engagement with politics. So, a lot of my key reference points have been urgent debates provoked by things like that.

> If there is one person that really influenced me early on, and this is a very strange person to cite, it is Louis Althusser. The most unpopular person you can possibly mention. (Bond and Featherstone 2009, 403)

Political issues were thus a primary source of Massey's intellectual inspiration and she described herself not as an academic, but as an intellectual, or better, as an 'intellectual activist'. After retiring in 2009 she commented that her greatest wish from then on was to do politics.

Talking about Massey's intellectual inspirations thus means, first and foremost, talking about her political imagination and engagement. From the events of 1968, when she was involved in the feminist movement as 'a reaction to the macho nature of 68' (Massey in Haesbaert et al. 2006), to the grassroots Occupy London movement, which rose against corporate greed in 2011, and the Kilburn Manifesto, which challenged the whole neoliberal order in her last few years, she was politically engaged. Her involvement ranged from that with the left wing of the British labour movement (left wing Labour Mayor of London Ken Livingstone was a friend from 1976 and wrote the foreword to her *Festschrift*; Featherstone and Painter 2013), to the Sandinista Revolution in Nicaragua, to post-apartheid struggles in South Africa, to the left Greek Syriza and to support for radical Hugo Chávez of Venezuela. Even on holiday in 2011 Massey chose to visit Cuba and a May Day tour organized by Cuban Solidarity Campaign (CSC), which promised a holiday with a difference, complete with opportunities to see every day Havana, and visits to workplaces, schools, local health centres and the May Day rally (see Figure 1.3).

Moreover, Massey's political stance governed her personal behaviour, whether her disdain for the University of Oxford's elitism ('the elitism of Oxford made me think that I didn't want to be an academic' (Massey in Engler 2012)) or her declining appointment as Officer of the Order of the British Empire (OBE). Her anti-academicism led her to avoid administrative positions and refuse invitations to leave the Open University – an innovative university founded in 1969 that focused on distance learning to bring higher education to those not served by traditional universities.

After graduating BA with a first class degree in geography in 1966 from the University of Oxford, Massey plunged into activism, especially the feminist cause (for example, through the group 'Women and socialism'). Between 1971 and 1972, and funded by a scholarship, she studied for a master's degree

Figure 1.3 Massey talking to staff and students in a Havana primary school during Cuba Solidarity Campaign's May Day Tour (2011). Photograph by Hilary Corton.

in regional science at the University of Pennsylvania, in the United States – joking that she wanted to 'know the enemy on the inside' (Albet and Benach 2012, 15). During this period and in reaction to her conventional education (revealed in her earlier writings and described rather hyperbolically by Barnes (2018, 126) as her 'dark past'), she took a course in French philosophy at the University of Pennsylvania focused on Louis Althusser – something she considered more important than regional science. She credited her ways of reinterpreting critical theory to these readings and experiences, focused on anti-essentialism and on the absence of a point of departure ('there is no point of departure', Massey in Featherstone and Painter 2013). Later, between 1980 and 1983, with Stuart Hall, Chantal Mouffe and Ernesto Laclau, she formed the Hegemony Group dedicated to rereading the writings of Antonio Gramsci and Louis Althusser, among others.

At the end of her master's course in the United States, Massey returned to London. Once in the UK she analysed, with Richard Meegan, the crisis of industrialization there and the regional inequalities it precipitated. She later wrote numerous texts about acute structural crises and consequent restructuring, developing a better understanding of economic geography, which, during the 1980s, she addressed through a trans-scalar analysis (Haesbaert 2013). She always used to think through scales, connecting different space–times. Massey linked specific aspects of the deindustrialization of Manchester with changes on a national and global level, something that is also revealed in the teaching materials produced by the Open University, for example, in her co-authored 1985 work *Changing Britain, Changing World: Geographical Perspectives.* Massey produced excellent teaching materials for her distance courses at the Open University whose very nature was a vehicle for her to teach her complex and creative concept of space, interactively. Her objective 'was not to theorize or even present case studies: it was about thinking through themes and concepts and presenting how they develop spatially'. Open University textbooks are completely renewed every six or seven years to give 'a collection of rigorous, but attractive and accessible works, exceptionally well-elaborated' (Albet and Benach 2012, 22).

Between the 1970s and 1980s Massey dealt with space not just as a surface of location, but as an inherent part of the logic of capitalist production. The inseparable link between space and power was an innovative interpretation at the time. She argued that space was central to understanding the social: neoclassical economics by contrast understood geography only as a locational factor, not a productive process. *Spatial Divisions of Labour* (1984 [Massey 1984a], 2nd edn 1995 [Massey 1995b]) was a noteworthy work on this subject.

With financing from the Social Science Research Council (SSRC) and in collaboration with Alejandrina Catalano, Massey completed another report on regional inequality, in land ownership in the UK. Her *Capital and Land* (1978), with its focus on land ownership and the theory of rent in the UK, attracted attention at the congress of the British Labour Party in 1973, though Massey was not a party member. The controversial book also influenced agrarian studies in Latin America. In 1986 Massey later wrote the script for and narrated the BBC–Open University documentary *Land use in Brazil* (1986) in the *Changing Britain, Changing World* series. She described land concentration in Brazil, especially in the Zona da Mata region in the state of Pernambuco, a well-chosen case study using which she analysed productive changes, correlating scales with a focus on the inseparability between space and time, and rural and urban.

Between 1968 and 1979 Massey worked at the Centre for Environmental Studies (CES), a research institute founded by Harold Wilson's second Labour government that produced influential research on planning and innovation. However, its autonomy and critical investigations irritated Margaret Thatcher's

Conservative government, which closed the centre, giving Massey first-hand experience of job loss in the Thatcher era (Featherstone 2016). Between 1981 and 1982 Massey completed a project at the London School of Economics that had started at CES and, with Michael Storper, Richard Walker and Michael Watts, she taught briefly at University of California, Berkeley, in 1981.

In 1982, after a period of instability in employment and having set her face against academia ('I thought that being an academic was something very elitist' (Massey in Engler 2012), Massey was elected Professor of Human Geography at the Open University (OU).

> At a certain point, the possibility of working at OU appeared and, this way, my problem was solved, personally because it was a university with a political project for popular and democratic education. It is the only place where I think I could be happy as an academic. It is a democratic project, you can go to OU without titles, but they offer the support necessary so that everyone can achieve a university level. It is an explicitly anti-elitist university. It is for the masses, it has thousands of students; in the college of Social Sciences, there must be more than ten thousand students. For me, this is how education should be, in different aspects, for anyone. In a course, I can have a wide range of people; there may be a diplomat, a housewife, a dockworker, and this fascinates me. They had already offered me work at elite universities, but I always continued working at OU. (Massey in Engler 2012)

Massey took up her position, despite not having a doctorate, and played a decisive role in transforming the Department of Geography in the Faculty of Social Sciences into one of the most important in the UK. According to her, 'It is a place which has a perhaps unique and certainly very collaborative and cooperative way of working. My colleagues ... have been a constant source of new ideas and of intellectual stimulation' (Massey 1995b, xiii). There she built very important intellectual partnerships, especially that with John Allen, with whom she organized several publications, and with other colleagues such as Allan Cochrane and Gillian Rose and, outside geography, with intellectuals like Stuart Hall.

Massey's long period at the Open University marked a career that allied the academic and the political and which reached audiences beyond academy, as she participated in radio and television programmes.

> During the 1990s, Massey visited Mexico City and Brazil ... to make documentary films for the BBC. The leading question throughout

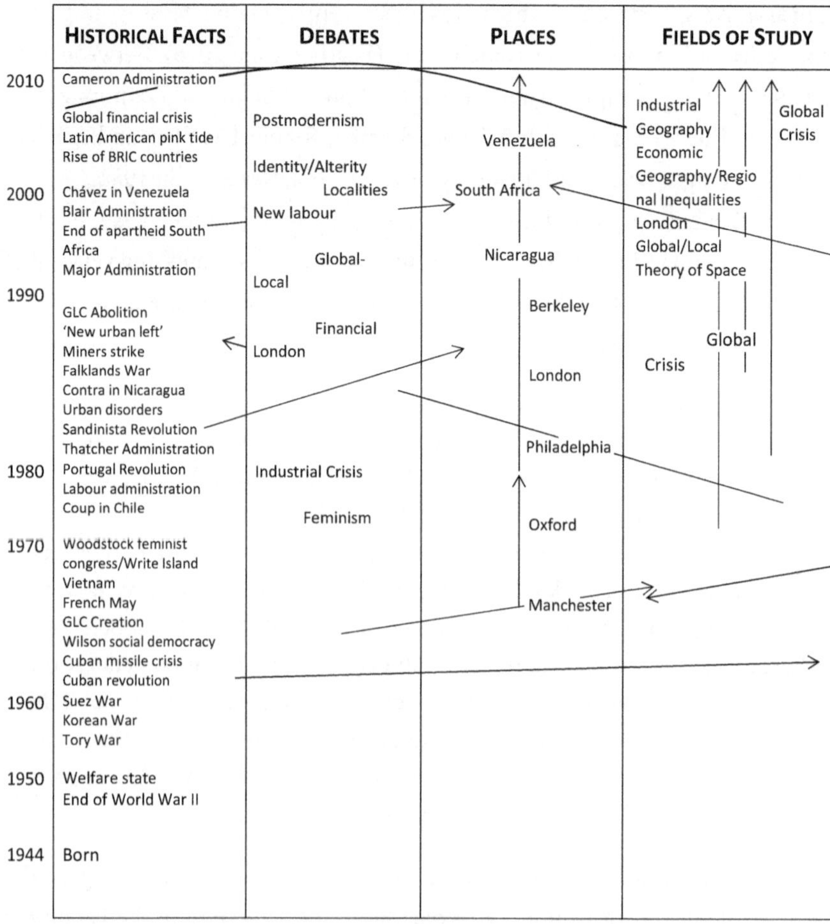

Table 1.1 Doreen Massey: summary of her intellectual changes in time and space (Albet and Benach 2012, 40–41). Reproduced with kind permission of Abel Albet and Núria Benach. It was originally published as 'Cuadro que sintetiza la evolución intelectual de Doreen Massey en el tempo y en el espacio'

> her journey across Mexico City was whose city is it? She interviewed different city inhabitants, from rural migrants living in deteriorated areas in historic neighbourhoods to advisors to transnational companies. Their answers led her to think that Mexico City [could not be understood solely] in light of the local context because all actors were involved in global networks. … Massey noticed that the gaps between the rich and the poor were becoming wider, at the same time that new forms of political and economic organization emerged that challenged the established order and privileges, opening possibilities of change for the future. (Zusman 2018, 576)

INFLUENCES	WORKS	STEPS	DISTRIBUTION
	'conjuncture series' World City	NLR SOUNDINGS	Robinson in Ruins Open Space
M. Gatens – G. Lloyd	For Space		European Social Forum
M. Bergson	Geographies of responsibility		Weather Project
	Human Geography Today Power geometries		The Dilapidated Dwelling
C. Mouffe E. Laclau		LSE OPEN UNIVERSITY	
S. Hall	Politics and space-time A global sense of place Flexible sexism		
		GLC-GLEB	BBC Programs
A. Gramsci	Spatial Divisions of Labour Geography Matters Anatomy of job loss		
L. Althusser K. Marx	In what sense a regional problem	C. ENV. STUDIES	OU Geography Manuals
	Capital and Land Towards a critique of location theory		

Table 1.1 (*Continued*)

As well as her television work, such as 'Imagination in the City' for BBC 4 (https://www.youtube.com/watch?v=936uuwkabf8) or 'Whose is the city?', a documentary about Mexico City (1999) (https://www.youtube.com/watch?v=936uuwkabf8), Massey was involved in cinema (for example, the documentary *Robinson in Ruins*; 2010, dir. Patrick Keiller) and projects at art galleries (for example, with sculptor and installation artist Ólafur Elíasson; and discussions at the Tate Gallery, London). From the start she contributed to geography journals, such as *Environment and Planning*, but also to periodicals with broader and more political focuses, such as *Marxism Today* and *Soundings*, which she founded with Stuart Hall and Michael Rustin, in 1995.

Massey was rare in combining an extraordinary academic career, constant political activism and teaching of the highest quality, and also in that most of her work was collaborative. Her sole authored books, though notable, are outnumbered by her many collaborative projects that she organized and in which she participated. The focus of her concern may be traced in her successive sole authored works: from capitalist spatial inequality and regional issues in *Spatial Divisions of Labour* (Massey 1984a) to the 'global' redefinition of place and the issue of gender in *Space, Place and Gender* (Massey 1994), to a new philosophy and politics of spatiality in *For Space* (Massey 2005). Critical analyses of *For Space* (Anderson 2008) and of *Spatial Divisions of Space* (Phelps 2008) show how innovative her work was.

Massey's intellectual work and her lived spaces were inseparable: her native industrial northern region inspired her works on regional inequality in England, while the Lake District features in *For Space* to exemplify the fluid nature of place. She had strong international ties, especially with Latin America, particularly Nicaragua, Mexico, Brazil and Venezuela (Zusman 2018; Ramírez Velázquez 2010). However, the space that most often inspired her work on place, globalization and neoliberalism was, without a doubt, London. Kilburn, in north London, was the basis for the discussion of the 'global sense of place' and used in the toponymous political manifesto written with Stuart Hall and Michael Rustin (2015), and London was the subject of her last sole authored book *World City* (Massey 2007).

Massey summarized her intellectual trajectory diagrammatically (Table 1.1) setting her intellectual concerns in geo-historical context and indicating her main influences (Albet and Benach 2012).

Intellectual legacy

Massey left a varied intellectual legacy. Theoretically, she exercised an important role in geography, and in the social sciences in a broader sense, by reinterpreting Marxism, establishing debates on gender and (re)elaborating post-structuralist critiques of space. Among her main conceptual contributions are those on the spatial division of labour, power geometries (of space–time compression), a global sense of place (the title of her most popular text) and space as 'the sphere of coexisting heterogeneity', 'in which distinct trajectories coexist' (Massey 2005, 9). In more recent works (Massey 2004), she also developed the idea of 'geographies of responsibility'. She always aimed to overcome dichotomies such as those between material and ideal, natural and social, fixity and mobility.

Massey's openness to different vehicles for the production and diffusion of knowledge made her uniquely well qualified to develop dialogues between different perspectives and authors.

> Apart from Marx and Althusser, Heidegger and Bergson, post-structuralism and feminism [to which we would add post-colonial thought], a last affinity of Massey's philosophical geography is the varied work that has taken place under the label 'complexity theory' ... Through the non-reductionist thinking of complexity theory Massey (2000b) inserts herself in a Spinozian tradition in which humans do not inhabit their own kingdom separate from the rest of nature. She also continues, though far from uncritically, geography's venerable tradition of conceptualizing the interaction between earth, culture and mobility. (Saldanha 2013, 50)

Although Massey worked in one of the very few British departments of geography dedicated exclusively to human geography, she aimed not to separate society from nature, as can be seen in her compelling critique of the conception of space as fixity. Over the course of her work, Massey incorporated critiques and broadened her conception of space and/or place. In 2000 she addressed the physical–human dichotomy in geography in 'Space–time, 'science' and the relationship between physical geography and human geography' (1999b), while in *For Space* (Massey 2005) she connected the natural and human dimensions of space and place, understanding the 'natural' as labile rather than rigid.

In her work on Marxism and spatial division of labour, Massey emphasized contingency, and granted primacy *a priori* to neither the global nor the local, and she developed the concept of the region–network that includes material and symbolic dimensions (Massey et al. 1998; and see Cochrane 2013). From early on she emphasized the importance of multidimensional and trans-scalar research, though she commented that she did not appreciate the concept of scale but worked much more on the intersections and the interconnections than on the delimitation and hierarchical overlapping of space. During the 1980s she reworked Althusserian Marxism in dialogue with Ernesto Laclau and Chantal Mouffe. Mouffe and Laclau (2005) coined the term 'radical democracy', arguing for the need to interpret democracy outside of the consensus model, and Massey emphasized the importance of 'thinking radical democracy spatially' (Massey 1995a).

Although she declined the label 'feminist geographer', Massey was one of the first authors to introduce the issue of gender in geography. In *Spatial Divisions of Labour* (Massey 1984b), for example, she addressed the substitution of full-time male labour by part-time, poorly paid female labour (Albet and Benach 2012), and in the second edition of the book she added two chapters and a section on 'Gender and feminism', in which she claims that, although the book was not 'written within what was at that time defined as the field of feminist geography', 'issues of "gender", in one way or another, run through the book', implying a mutual connection between relations of production and gender (that is, capitalism and the patriarchy) (1995, 343). Massey was nonetheless cautious about having done 'gender Geography', claiming that, despite having initially included Marxism,

> I didn't know how to include feminism. … I mentioned women, I made rude remarks about men, I did all those classic things, but I didn't put a theorized feminism into my work. … I knew the category itself was problematical. Although I was willing to write about that, and you see lots of reference here and there, I didn't know how to get a purchase on it intellectually. … So I was incorporating the Marxism but not the feminism, and it would be a long time before I incorporated the feminism, even though I was politically active as a feminist from the late 1960s. (Freytag and Hoyler 1999, 87)

Massey also addressed 'the type of political identity of women' but did not see the need to 'dedicate all of our efforts to prioritizing the topic of gender'. What she tried to do was 'write about any topic as someone who is against sexism and against heteronormativity'. This does not mean, however, 'that I do not believe that other people should be dedicated to it' (Albet and Benach 2012, 54). As Silva, Ornat and Chimin Jr. (2017) describe it: 'There is without a doubt a confusion between studying gender in geography and producing feminist geographies. Doreen Massey was very clear on this aspect. For her, being a feminist geographer does not directly refer to the topic of gender and addressing gender in geographical research is not directly linked to doing feminist science. Feminism in academic practice is a broader subversive perspective and, in her view, it was possible to be a feminist geographer without addressing gender, per se' (2017, 12). One of her more emphatic and best-known texts is 'Flexible sexism' (1991, reprinted in *Space, Place and Gender* (Massey 1994)), in which she promotes a compelling dialogue with two notable geographical works on postmodernity: Edward Soja's *Postmodern Geographies* (1989) and David Harvey's *Condition of Postmodernity* (1989). Condemning the absence of the feminist contribution in those works and positioning herself in relation to the debate on postmodernity, she claims:

> I have great sympathy with the overall projects of both these books. Soja is struggling to be postmodern but really remaining in many ways modern. Harvey is quite clear for modernism but wanting, he says, to change it in ways which will respond to certain inadequacies. I too, I would like to retain strong aspects of what characterizes the modernist project, most particularly its commitment to change, hopefully progressive. I also agree strongly with Harvey's defence of much of what has been achieved in its name. But it is necessary also to recognize the inadequacies of the modernist project in its dominant form. … The answers which most postmodernism has so far provided may well be mistaken, but the challenges it poses must surely be addressed. Moreover, one stream of thought which has been raising many of the same issues for far longer … is feminism. Even so, potential contributions of feminism have simply been ignored. (Massey 1994, 238–9)

Among Massey's important contributions were the concepts of the 'global sense of place' and of 'power geometries' of space–time compression. Massey simultaneously recognized the importance of, and questioned the limits of, Harvey's proposals (Cresswell 2004), critiquing the idea of space–time compression when expressed only through socio-economic, class conditions by considering other factors such as gender, ethnicity, religion, physical abilities and age or generational group, all of which have important implications for the multiple geometries with which power is established in global processes that compress our relations in time and space. She bases place as a meeting place not on the characteristics of the, often globalized, phenomena but on the way in which such phenomena combine or connect in each place.

Massey remarked that her notion of power geometries had been very intuitive and that, until it was taken up by Hugo Chávez in Venezuela, she had not effectively deepened it as a concept.

> Not exactly a concept; it is a kind of provocation to imagine, to stimulate the imagination. Moreover, geometry sounds very rigid and, this is why people say 'maybe it should be power topology' (the title of a book by her colleague John Allen), or something like that, and I totally agree. I do not refer to geometry in the sense that there are rigid Pythagorean triangles everywhere, and the idea of topology is suggestive, because it talks about the possibility of change and the mutability of space, which is precisely what I am saying. (Román and García 2008, 335)

To her surprise, Chavismo included power geometries as one of the five motors of the 'Bolivarian revolution'. On more than one occasion she was invited to Venezuela to discuss the topic and to become familiar with the experiments to produce more geographical egalitarianism based on these 'new power geometries'. A conference at the Seminar of Dirección de Cooperación Internacional in Caracas was dedicated to exploring power geometries (Conferencia 3: Las geometrías del poder en el contexto de la reforma constitucional – Conference 3: The power geometries in the context of constitutional reform). In her conference paper she highlights that power geometries are a movement to 'give a voice to all places', particularly via local, everyday self-management of 'consejos comunales' (communal councils): 'Before the poor neighbourhoods were excluded from the political debate; they did not have a political voice and I believe that they do now' (Massey in Engler 2012).

In an interpretation clearly (though not explicitly) inspired by Foucault, Massey discusses the relational sense of power as a practice and exercise, disseminated between social relations, and claims: 'Power always has a cartography; you can make maps of social, political and economic power' (Zusman 2018, 577). In lectures and conferences in Venezuela, which she visited in 2008 to advise on and discuss the concept of power geometries, she summarized her reflections on the power–geometries of space–time thus:

a) it is 'a general concept, a way of understanding human Geography', without implying 'any specific form of geometry', each political dispute producing a complex, concrete geometry;
b) it implies a concept of relational power, inherent to entities (territories, social forces, communities), which, in turn, are mutually constituted through these relations;
c) they are always involved in time and, therefore, subject to transformation;
d) they exist in all spheres – not only political, but also economic, cultural, educational, etc., and on all levels – from local to global; and
e) they always imply the presence of conflict, with power relations where it will always be necessary to negotiate.

In short, she aimed with this concept – and based on her Venezuelan experience – to insist on the political content of space and the fact that power itself has a spatial form (a geometry), not limited to territorial definitions, though they are important as a means of pursuing the fundamental end of being able to redistribute social power.

Finally, the relational concept of space for Massey centered on overcoming space–time essentialism and division, aimed to praise its openness, multiplicity and contingency, while also highlighting its link to broader circuits of power. Her attention to many subjectivities and her deep concern with social injustice always went hand in hand as part of a radically democratic political project (Massey 1995a).

Her essays and articles from the 1990s and the 2000s as well as in *For Space* (Massey 2005) show Massey's political intent in her approach to space/spatiality, terms she used and defined as 'interchangeable'. Whether noun, adjective or verb, space/spatiality/spatial/spatialize are concepts that evoke and ritualize the coevalness of life, aside from attempts or strategies of capture, closure and immobility. Massey (2005) takes up Johannes Fabian's idea of 'coevalness [which] aims at recognizing cotemporality as the condition for truly dialectical confrontation' (Fabian in Massey 2005, 69). She considered it a strategy for claiming the inseparability of space–time and, as she said:

> Space is the sphere of the possibility of the existence of multiplicity, it is the sphere in which distinct trajectories coexist, it is the sphere of the possibility of the existence of more than one voice. Without space, no multiplicity, without multiplicity no space. If space is indeed the product of interrelations, then it must be predicated upon the existence of plurality. Multiplicity and space are co-constitutive. (1999, 28)

The guarantee that there is no multiplicity without spatiality, and vice-versa, allied to the ideas of juxtaposition and incompleteness, presents links that bring her work closer to post-structuralism – 'closer' because she does not deny the contribution of structuralist perspectives to social thought. In the chapter of *For Space* ironically entitled 'The prison-house of synchrony' Massey considers that the conjunction between structures and social thought gave scientific status to geography. Citing Soja, she reports that structuralism had been 'one of the 20th century's most important avenues for the reassertion of space in critical social theory' (Soja 1989 cited in Massey 2005, 37).

The adherence to anti-essentialism, claims Massey, is dependent on a different spatial political imagination. For this reason we must question the centrality of the subject and recognize the incompleteness of identity politics, and, rather, emphasize that space participates in the establishment of political subjectivities (2005, 42). In Massey's words, 'We should be wary about claims to authenticity based on notions of unchanging identity' (Massey 1999a, 29). She argues that space can be conceived as 'open, unfinished, always in transformation'.

Massey's insistence that space and/or place are always open, fluid and interconnected was not universally accepted. Some authors criticized her for an 'orthodoxy of the contingent and the open' (Callard 2005, 224), and this prompted a rich debate between those advocating progressive and conservative senses of place. As Cresswell (2004) claimed, commenting on a work by May (1996), there are 'multiple ways in which people relate to the same place'.

> Issues of boundaries and rootedness and connections are still there but they are used in complicated ways by people. The simple, observable, fact of diversity does not necessarily produce a progressive sense of place and the search for roots in history does not have to be reactionary. (Cresswell 2004, 79)

Nevertheless, Massey was decisive in reformulating the debate and reintroducing to geography a profoundly innovative approach to places 'not as points or areas on maps, but as integrations of space and time; as spatio-temporal events', collections of trajectories, 'a moment within power-geometries of space-time' (Massey 2005, 130–31).

Later life, recognition and legacy

Massey was from early on widely recognized as a major voice in geography: for example, she was awarded the Vautrin Lud Prize, in France, in 1998 (the most important award in geography); she was invited to give the annual Hettner lecture series at the University of Heidelberg's Geographisches Institut (1998), publishing her work the following year as 'Power-geometries and the politics of space-time' (1999). In later life, richly deserved honours came thick and fast: honorary doctorates from Queen Mary University of London, Edinburgh and Glasgow; the National University of Ireland, Zürich and Harokopio; prizes and medals from the American Association of Geographers, the Royal Geographical Society (the Victoria Medal, in 1994), the Scottish Geographical Society (Centenary Medal, in 2003) and the Swedish Society of Anthropology and Geography (Anders Retzius Medal, in 2003). She was elected fellow of the Academy of Social Sciences (1999), the Royal Society of Arts and the British Academy (2002). In 2009 her retired life was marked by the granting of the title Professor Emerita from the Open University and a Royal Geographical Society roundtable 'Spatial Delights: an engagement with the work of Doreen Massey' (2009). Her work was discussed, for example, in *Spatial Politics: Essays for Doreen Massey* (2013), organized by David Featherstone and Joe Painter, and *Doreen Massey: un sentido global de lugar*, organized by Abel Albet and Núria Benach in Barcelona (Albet and Benach 2012); after her death *The Doreen Massey Reader* (Christophers et al. 2018); *Doreen Massey: Critical Dialogues* (Peck et al. 2018), edited by Brett Christophers, Rebecca Lave, Jamie Peck and Marion Werner; and the Doreen Massey dossier in the Brazilian journal GEOgraphia (in 2017).

Doreen Massey died suddenly and prematurely on 11 March 2016 at her home in Ariel Road, Kilburn, London. Reaction to her death was heartfelt and the tributes that followed often very personal. Gillian Rose, for example, her colleague at the Open University, recalled:

> She wasn't always an easy person to work with. She could be very critical; she could insist on things being done her way; she didn't like any kind of admin. She could also, far more often, be incredibly warm – to everyone and anyone, absolutely – and she was one of the most charismatic speakers I have ever heard. I remember her tiny frame absolutely filling one enormous lecture hall with energy and passion, extemporising from handwritten notes, intensifying the entire space. (Rose 2016)

Hilary Wainwright, another colleague, remembered that 'Doreen's body could sometimes be frail but her mind was brilliant and tough. As a character she was strong, passionate, curious and also imbued of modesty and kindness. She radiated political energy and humanity, sparking with a cheeky wit' (Wainwright 2016).

In addition to a circle of grieving friends, Massey left a vast intellectual, pedagogical and political legacy, but also her profound love for the world and for life. In the revealing interview with Albet and Benach (2012), when questioned about her role as 'one of the most significant and well-respected voices in social sciences' (with her work translated into languages such as Greek, Japanese, Polish, German, French, Spanish and Portuguese), Massey remarked that one of the most curious and ironic consequences of postmodernism and post-structuralism, which, it was alleged, would bring about the death of geography, was the emergence of 10 to 20 geographers as 'big stars' in global academia, a group to which, she claimed, 'I sincerely don't feel like I belong.' She added:

> And I don't want to convert into one of them ... because I want to have my own life ... I want to walk in the country and take care of my garden, watch football games and observe birds in the wild, etc. Politics also occupies a lot of my time and I like my work, but I am not prepared to dedicate 24 hours a day to geography. ... Sometimes I ask myself: 'could I have a more important voice?' I look back and think: 'David Harvey is everywhere: why don't I do the same thing?' Well, because I am not a man: maybe this makes things different. (Albet and Benach 2012, 49–50)

Doreen Massey was inimitable. She brought perspectives never before developed, combining broad theoretical–philosophical reflection with political propositions, and she knew how to read the world with her own eyes, a kind of 'geographic intuition', a way of looking at the world through space – something many geographers claim but few truly possess. Breaking – or, in her words – 'negotiating' disciplinary boundaries (Massey 1999e), she always maintained the geographical consistency of her thought and integrated everyday experiences with complex theory to elucidate key questions such as the sense of place and the inseparability of space and time.

Massey boldly developed a creative philosophy of space, and based on her own experiences, created a unique language for interpreting and building space. Perhaps this is a way of summarizing the 'Massey style' of describing and communicating, as Albet and Benach (2012) claim, in texts, video, on the radio, in classes, in film and through other art forms. Massey amplified the hopes emerging from processes of resistance to authoritarianism, such as during the liberal reaction to the bailout of financial institutions after the financial 2008 crash. The resistance started with her everyday actions and with her small fragile body, through multiple *r–existires*, in the term coined by C.W. Porto-Gonçalves (2001) to convey the idea that a combination of resistance and existence is the founding principle of social movements.

Massey's life path dissolved frontiers between personal life, political activism and intellectual reflection. Her refusal to conform exemplified the fact that thinking about space today is an integrated and radical project of democracy. From the Greater London Council to the Kilburn Manifesto, from Occupy London to the *Consejos Barriales* (Neighborhood Councils) of Caracas, Massey inspired, encouraged and designed multiple geographies ('she geo-graphed', in the sense of 'writing' the earth), translating in a unique way the coevalness and the polyphonic synchrony of places. This is why, for her, 'the challenge of space is precisely the challenge of the full recognition of coeval others' (Massey 2007, 210).

Acknowledgements

We extend warm thanks to Hilary Corton, Doreen Massey's sister for her generous contribution to, and attentive reading of, this essay and for providing one of the photographs included in it, John Allen, Allan Cochrane and Jan Smith from the Open University; Abel Albet and Núria Benach from the Universidad Autónoma and Universidad de Barcelona; Perla Zusman, from the Universidad de Buenos Aires; Antonio Hernandez, from British Library. Our special thanks to André Novaes and Elizabeth Baigent who have read this manuscript and made valuable suggestions for its improvement.

Works about Doreen Massey

Anderson, B. (2008), 'For Space (2005): Doreen Massey', in P. Hubbard, R. Kitchin and G. Valentine (eds), *Key Texts in Human Geography*. London: Sage, 225–33.

[Association of American Geographers] (2016), 'In Memorium: Doreen Massey', http: //news.aag.org/2016/03/in-memoriam-doreen-massey/ (accessed August 2018).

Barnes, T. (2018), 'Her dark past', in M. J. Werner, J. Peck, R. Lave and B. Christophers (eds), *Doreen Massey: Critical Dialogues*. Newcastle-upon-Tyne: Agenda, 124–45.

Callard, F. (2005), 'Doreen Massey', in P. Hubbard et al. (eds), *Key Thinkers on Space and Place*. London, Thousand Oaks, New Delhi: Sage, 299–306.

Castree, N. (2016), 'A tribute to Doreen Massey', *Progress in Human Geography* 40(5), 585–92.

Christophers, B., Lave, R., Peck, J. and Werner, M., eds (2018), *The Doreen Massey Reader*. Newcastle-upon-Tyne: Agenda.

Clarke, J. (2016), 'Doreen Massey (1944–2016): Making geography matter', *Cultural Studies* 30, 357–61.

Cochrane, A. (2013), 'Spatial divisions and regional assemblages', in D. Featherstone and J. Painter (eds), *Spatial Politics: Essays for Doreen Massey*. Chichester: Wiley–Blackwell, 87–98.

Cresswell, T. (2004), *Place: A Short Introduction*. Oxford: Blackwell.

Elden, S., ed. (2016), 'Progressive geographies', https://progressivegeographies.com/2016/03/13/tributes-to-doreen-massey/ (accessed August 2018).

Featherstone, D. (2016), 'Obituary', *Guardian*, 27 March 2016.

Featherstone, D. and Painter, J. (2013), *Spatial Politics: Essays for Doreen Massey*. Chichester: Wiley–Blackwell.

Haesbaert, R. (2013), 'A global sense of place and multi-territoriality: Notes for dialogue from a "peripheral" point of view', in D. Featherstone and J. Painter (eds), *Spatial Politics: Essays for Doreen Massey*. Chichester: Wiley–Blackwell, 146 57.

https://www.manchestereveningnews.co.uk/news/greater-manchester-news/tributes-after-death-geographer-acclaimed-11033331 (accessed August 2018).

http://www.scielo.org.mx/scielo.php?script=sci_arttext&pid=S0188-46112010000200015&lng=es&nrm=iso (accessed August 2018).

Kitchin, R. (2016), 'Geographers matter! Doreen Massey (1944–2016)', *Social and Cultural Geography* 17, 813–17.

Laclau, E. (2005), *Nuevas reflexiones sobre la revolución de nuestro tiempo*. Buenos Aires: Nueva Visión.

Lee, R. (2016), 'Doreen Barbara Massey', *Biographical Memoirs of Fellows of the British Academy* 17, 145–78.

Littler, J. and Gilbert, J. (2016), 'The Doreen Massey we knew', Open Democracy [blog] https://www.opendemocracy.net/uk/jeremy-gilbert-jo-littler/doreen-massey-we-knew (accessed August 2018).

Painter, J. (2016), 'Doreen Massey (1944–2016): An appreciation', *Soundings* 63, 38–44.

Peck, J., Werner, M., Lave, R. and Christophers, B. (2018), 'Out of place – Doreen Massey, radical geographer', in J. Peck, M. Werner, R. Lave, B. Christophers (eds), *Doreen Massey: Critical Dialogues*. Newcastle-upon-Tyne: Agenda, 32–102.

Phelps, N. (2008), '*Spatial Divisions of Labour* (1984): Doreen Massey', in P. Hubbard, R. Kitchin and G. Valentine (eds), *Key Texts in Human Geography*. London: Sage, 83–90.

Philippopoulos-Mihalopoulos, A. (2016), 'Doreen Massey has died', *Critical Legal Thinking* [blog] https://criticallegalthinking.com/author/andreas-philippopoulos-mihalopoulos/ accessed August 2018.

Ramírez Velázquez, B. (2010), 'Presentación y Mesa Redonda: Doreen Massey y las "geometrías de poder"'. *Investigaciones Geográficas* 72. México. August 2010.

Rose, G. (2016), 'Remembering Doreen', https://visualmetheodculture.wordpress.com/2016/03/12/remembering-doreen/ (accessed August 2018).

Saldanha, A. (2013), 'Power-geometry as philosophy of space', in D. Featherstone and J. Painter (eds), *Spatial Politics: Essays for Doreen Massey*. Chichester: Wiley–Blackwell, 44–55.

Sevillha-Buitrago, A. (2016), 'Fallece Doreen Massey', *Multipliciudades*, https://multipliciudades.org/2016/03/12/fallece-doreen-massey/ (accessed August 2018).

Silva, J., Ornat, M. and Chimin, Jr, B. (2017), '"Não me chame de Senhora, eu sou feminista!" Posicionalidade e reflexibilidade na produção geográfica de Doreen Massey', *GEOgraphia* 40, http://periodicos.uff.br/geographia/issue/view/857 (accessed 20 March 2018).

Skelton, T. (2016), 'In memorium: Doreen Massey (1944–2016)', in *Society and Space*, http://societyandspace.org/2016/05/10/doreen-massey-in-memoriam/ open site (accessed 20 March 2018).

Wainwright, H. (2016), '"How we will miss that chuckle": My friend, Doreen Massey', *Open Democracy*, https://www.opendemocracy.net/uk/hilary-wainwright/how-we-will-miss-that-chuckle-my-friend-doreen-massey (accessed 20 March 2018).

Werner, M.J., Peck, J., Lave, R. and Christophers, B., eds (2018), *Doreen Massey: Critical Dialogues*. Newcastle-upon-Tyne: Agenda.

Zusman, P. (2018), 'Doreen Massey and Latin America', in M. J. Werner, J. Peck, R. Lave and B. Christophers (eds), *Doreen Massey: Critical Dialogues*. Newcastle-upon-Tyne: Agenda, 303–12.

Selected publications by Doreen Massey

1973 'Towards a critique of industrial location theory', *Antipode* 5, 33–9.

1974 '*Social Justice and the City*: A review', *Environment and Planning A* 6(2), 229–35.

1978a 'Regionalism: Some current issues', *Capital and Class* 6, 106–25.

1978b with Catalano, A., *Capital and Land: Landwnership by Capital in Great Britain*. London: Edward Arnold.

1979a with Meegan, R., 'The geography of industrial reorganisation: The spatial effects of the restructuring of the electrical engineering sector under the Industrial Reorganisation Corporation', *Progress in Planning* 10(3), 155–237.

1979b 'In what sense a regional problem?', *Regional Studies* 13, 233–43.

1982 with Meegan, R., *The Anatomy of Job Loss: The How, Why and Where of Employment Decline*. London: Methuen.

1983 'Industrial restructuring as class restructuring: production decentralization and local uniqueness', *Regional Studies* 17(2), 73–89.

1984a *Spatial Divisions of Labour: Social Structures and the Geography of Production*. Basingstoke: Macmillan.

1984b ed. with Allen, J., *Geography Matters: A Reader*. Cambridge: Cambridge University Press.

1984c 'Introduction: Geography matters', in D. Massey and J. Allen (eds), *Geography Matters! A Reader*. Cambridge: Cmabridge University Press.

1985a 'New directions in space', in D. Gregory and J. Urry (eds), *Social Relations and Spatial Structures*. London: Macmillan, 9–19.

1985b with Wainwright, H., 'Beyond the coalfields: The work of the miners' support groups', in H. Beynon (ed.), *Digging Deeper: Issues in the Miners' Strike*. London: Verso, 149–68.

1986 'Nicaragua: Some reflections on socio-spatial issues in a society in transition', *Antipode* 18(3), 322–31.

1988a 'Uneven development: Social change and spatial divisions of labour', in D. Massey and J. Allen (eds), *Uneven Re-Development: Cities and Regions in Transition: A Reader*. London: Hodder and Stoughton, 250–76.

1988b ed. with Allen, J., *The Economy in Question: Restructuring Britain*. London: Sage.

1991a 'The political place of locality studies', *Environment and Planning D: Society and Space* 23, 267–81.

1991b 'A global sense of place', *Marxism Today* (June), 24–9.

1991c 'Flexible sexism', *Environment and Planning D: Society and Space* 9, 31–57.

1991d with Quintas, P. and Wield, D., *High-Tech Fantasies: Science Parks in Society, Science and Space*. London: Routledge.

1992a 'A place called home?' *New Formations* 17, 3–15.

1992b 'Politics and space/time', *New Left Review* 196, 65–84.

1994 *Space, Place and Gender*. Cambridge: Polity Press.

1995a 'Thinking radical democracy spatially', *Environment and Planning D: Society and Space* 13, 283–8.

1995b *Spatial Divisions of Labour: Social Structures and the Geography of Production*, 2nd edn. London: Macmillan.

1997 'Spatial disruptions', in S. Golding (ed.), *The Eight Technologies of Otherness*. London: Routledge.

1998 with Allen, J., Cochrane, A., Charlesworth, J., Court, G., Henry, N. and Sarre, P. *Rethinking the Region*. London: Routledge.

1999a *Power-Geometries and the Politics of Space-Time*. Heidelberg: Department of Geography, University of Heidelberg.

1999b 'Space–time, "science" and the relationship between physical geography and human geography', *Transactions of the Institute of British Geographers* 24, 261–76.

1999c 'Imagining globalization: Power–geometries of time–space', in A. Brah, M. J. Hickman and M. Mac an Ghaill (eds), *Global Futures: Migration, Environment and Globalization*. Basingstoke: Macmillan, 27–44.

1999d 'Philosophy and politics of spatiality: Some considerations', in D. Massey *Power–Geometries and the Politics of Space–Time*. Heidelberg: Department of Geography, University of Heidelberg., 27–42.

1999e 'Negotiating disciplinary boundaries', *Current Sociology* 47(4), 5–12.

2001 'Living in Wythenshawe', in I. Borden, J. Kerr, J. Rendell and A. Pivaro (eds), *The Unknown City: Contesting Architecture and Social Space*. Cambridge: MIT Press, 458–75.

2003a 'Some times of space', in S. May (ed.), *Olafur Eliasson: The Weather Project*. London: Tate Publishing, 107–18.

2003b with Amin, A. and Thrift, N., *Decentering the Nation: A Radical Approach to Regional Inequality*. London: Catalyst.

2004 'Geographies of responsibility', *Geografiska Annaler B* 86(1), 5–18.

2005 *For Space*. London: Sage.

2006 'Landscape as a provocation: Reflections on moving mountains', *Journal of Material Culture* 11(1/2), 33–48.

2007 *World City*. Cambridge: Polity.

2008a 'When theory meets politics', *Antipode* 40(3), 492–7.

2008b 'Geometrías intenacionales del poder y la política de uma "ciudad global": pensamentos desde Londres', [lecture at CEDES] *Cuadernos CEDES*, 25/68. Caracas, August.

2009 'Invention and hard work', in J. Pugh (ed.), *What is Radical Politics Today?* Basingstoke: Palgrave Macmillan, 136–42.

2010 *World City*, 2nd edn. Cambridge: Polity.

2011 'A counterhegemonic relationality of place', in E. McCann and K. Ward (eds), *Mobile Urbanism: Cities and Policy Making in the Global Age*. Minneapolis: University of Minnesota Press, 1–14.

2014 'The Kilburn Manifesto: After neoliberalism?', *Environment and Planning A* 46(9), 2034–41.

2015a 'Globalización, espacio y poder', in *CEPAL Memoria del Primer Encuentro de Expertos Gubernamentales em Políticas de Desarrollo Territorial em América Latina y el Caribe*. Santiago de Chile: CEPAL, Naciones Unidas, 9–14.

2015b with Hall, S. and Rustin M., *After Neoliberalism: The Kilburn Manifesto*. London: Lawrence and Wishart.

Interviews with Doreen Massey

Albet, A. and Benach, N. (2012), 'Conversando con Doreen Massey: una intelectual comprometida con su "espacio/tiempo"', in A. Albet and N. Benach (eds), *Doreen Massey: un sentido global de lugar*. Barcelona: Icaria.

Bond, S. and Featherstone, D. (2009), 'The possibilities of a politics of place beyond place? A conversation with Doreen Massey', *Scottish Geographical Journal* 125(3–4), 401–20.

Engler, V. (2012), *Página 12*, 29 October 2012, https://www.pagina12.com.ar/diario/dialogos/21-206595-2012-10-29.html (accessed August 2018).

Featherstone, D., et al. (2013), '"Stories so far": A conversation with Doreen Massey', in D. Featherstone and J. Painter (eds), *Spatial Politics: Essays for Doreen Massey*. Chichester: Wiley–Blackwell.

Freytag, T. and Hoyler, M. (1999), '"I feel as if I've been able to reinvent myself": A biographical interview with Doreen Massey', in D. Massey, *Power-Geometries and the Politics of Space–Time (Hettner Lecture 1998)*. Heidelberg: Department of Geography, University of Heidelberg.

Haesbaert, R. Scheibe, L. F., Arruda, S. M. and Buss, M. D. (2006), *Geosul*, Florianópolis, 21, 42, July/December, https://periodicos.ufsc.br/index.php/geosul/article/view/12869 (accessed 15 March 2018).

Massey, D. and Stevens, A. (2010), 'The future of landscape: Doreen Massey: An interview by A. Stevens', *3:AM Magazine*, 20 September 2010, www.3ammagazine.com/3am/the-future-of-landscape-doreen-massey (accessed 8 January 2018).

Román-Velázquez, P. and García-Vargas, A. (2008), 'Hay que traer el espacio a la vida: entrevista con Doreen Massey'. *Signo y Pensamiento* 27(53), 328–43.

[Social Science Space] (2016), 'The geographer of space and power: Doreen Massey (1944–2016)', https://www.socialsciencespace.com/2016/03/the-geographer-of-space-and-power-doreen-massey-1944-2016/ (accessed 8 January 2018).

Warburton, N. (2013), 'Doreen Massey on space', *Space Social Science Bites* (blog), www.socialsciencebites.com (accessed 8 January 2018).

Films/videos/podcast with Doreen Massey

'After Neo-liberalism? The Kilburn Manifesto', *Taking Soundings* (Leeds, 19 June 2013), https://www.youtube.com/watch?v=P2DyPUDzXIM (accessed 14 February 2019).

Chanan, M. (dir.) and Salter, L. (writer) (2012), *Secret City*, http://secretcity-thefilm.com/the-film/ (accessed 11 February 2018).

Debates y Combates (3rd edn., Tecnópolis, Argentina, 11 October 2012), https://www.youtube.com/watch?v=9wn4QphR4Yw (accessed 14 February 2019).

IV Seminario Atlántico de Pensamiento (Spain, 11 March 2011), https://www.youtube.com/watch?v=uBWWm-NINow (accessed 15 February 2018).

Land use in Brazil. (1986), Dir. Hugh Phillips, Doreen Massey (BBC/Open University), UK.

Massey, D. and Mouffe, C. (2011), 'Spatial justice introduction', 19 November 2011, https://www.westminster.ac.uk/news-and-events/events/spatial-justice-radical-foundations (accessed 14 February 2019).

Progressive London: The Global Economic Crisis Conference (13 June 2009), https://www.youtube.com/watch?v=2el0ZBHccqw (accessed 14 February 2019).

Other references

Corton, H. (2018), Personal Comments to Rogério Haesbaert.

Harvey, D. (1989), *The Condition of Postmodernity: An Enquiry into the Origins of Cultural Change*. Oxford: Wiley-Blackwell.

May, J. (1996), 'Globalization and the politics of place: Place and identity in an inner London neighbourhood', *Transactions of the Institute of British Geographers* 21(1), 194–215.

Porto-Gonçalves, C. W. (2001), *Amazônia, Amazônias*. São Paulo: Contexto.

Robinson in Ruins, (2010), Dir. Patrick Keiller (Documentary, UK).

Soja, E. (1989), *Postmodern Geographies. The Reassertion of Space in Critical Social Theory*. New York: Verso.

Chronology

1944	Born in Wythenshawe, Manchester, 3 January
1955–62	Studies at Manchester High School for Girls
1963–1966	Reads for BA in geography at St Hugh's College, University of Oxford as Irene Shrigley scholar
1966	Special college prize on her graduation first class
1968	Enters the Centre for Environmental Studies (CES)
1971–1972	studies for MA in regional science, University of Pennsylvania
1978	Publishes, with Alejandrina Catalano, *Capital and Land*
1981	three months' visiting professorship in the Department of Geography at the University of California, Berkeley

1982	Publishes, with Richard Meegan, *Anatomy of Job Loss*
1982	Appointed Professor of Geography at the Open University
1984	Publishes *Spatial Divisions of Labour* and, with John Allen, *Geography Matters*
1985–1986	Sabbatical at the Instituto Nicaraguense de Investigaciones Económicas y Sociales (INIES), Managua, Nicaragua
1991	Publishes 'A global sense of place' in *Marxism Today*
1991	Publishes 'Flexible sexism' in *Environment and Planning D: Society and Space*
1994	Receives the Victoria Medal of the Royal Geographical Society
1994	Publishes *Space, Place and Gender*
1995	Creates, with Stuart Hall and Michael Rustin the journal *Soundings*
1998	Invited to give the Hettner Lecture in Heidelberg
1998	Awarded the Prix Vautrin Lud in Saint-Dieu-des-Vosges, France
1999	Elected Fellow of the Academy of Social Sciences
2000	Elected Fellow of the Royal Society of Arts
2002	Elected Fellow of the British Academy
2003	Receives the gold Anders Retzius Medal from the Swedish Society for Anthropology and Geography
2003	Receives the Centenary Medal of Royal Scottish Geographical Society
2005	Publishes *For Space*
2006	Receives an honorary doctorate from the University of Edinburgh, Scotland
2006	Receives an honorary doctorate from the National University of Ireland
2007	Invited by the Venezuelan government to discuss her concept of power–geometries, considered one of the pillars of the Hugo Chávez political project
2007	Publishes *World City* (2nd edn 2010)
2009	Retires from the Open University with title Professor Emerita
2009	Receives an honorary doctorate from the University of Glasgow, Scotland
2010	Receives an honorary doctorate from Queen Mary University of London
2012	Receives an honorary doctorate from Harokopio University, Greece
2013	Receives an honorary doctorate from the University of Zürich, Switzerland
2014	Receives the Presidential Achievement Award from the Association of American Geographers
2016	Dies at home in Ariel Road, Kilburn, north London, 11 March

2

Women geographers at the University of Oxford

Elizabeth Baigent, Claire Hann, Susan Squibb
and Heather Viles

'To educate young women like young men and with young men, a thing inexpedient and immodest': thus J. W Burgon's University Sermon of 8 June 1884 epitomized the practical, social and moral opposition to women's education at the University of Oxford. He sorrowfully admitted in that sermon that the battle for educational *apartheid* had in his view already been lost, and geography presented just what Burgon feared when, three years later, women and men began studying the subject at Oxford, sharing substantially the same curriculum and being taught side by side in the lecture hall, and before long in the laboratory and in the field.

The histories of geography at Oxford and of women at Oxford have interesting parallels. Both struggled for legitimacy in an ancient University of resident men whose curriculum was highly traditional and which was socially conservative. Geographers and women were both outsiders, and, rather than storming the bastion of the collegiate University's structures, geographers and women encamped outside it and gained entrance by gradually undermining what some thought its crucial foundations: its eschewing of vocational subjects, its maleness and, to a lesser extent, its role as the 'natural' destination of public school men. The same incremental approach underscored both the strategy of the first Reader in geography who wanted gradually 'to infuse Geography into the general life of Oxford' (cited in Firth 1918, 11) and the tactics of women at Oxford who obtained, in the words of pioneer Annie Rogers, *Degrees by Degrees* (1938).

While similarity between the histories of Oxford's geographers and Oxford's women has thus far been largely lost on historians of geography at Oxford, the discipline's history is told in clearly gendered terms. Geography is described in feminine terms when it languishes. It is a 'Cinderella' (Heyes 1886, 7; Myers 1929, 456; Crawford 1955, 69) and there is no better evidence of its plight than its dependence on women – hence the, surely apocryphal, stories that an early geography lecture depended on a map drawn by 'two ladies' using the laundry whitener 'Reckitt's blue' (Heyes 1886, 6); and that two of the three attenders at the first lecture by the Reader in geography in 1887 were ladies who brought their knitting (Mackinder 1931, 321; Coones 1989, 16; were they perhaps chaperones?). While geography's vulnerability at Oxford is couched in feminine terms, its historians describe its successes in masculine language. Carlylean accounts describe a school made and then saved from possible demise by Great Men: Halford Mackinder (1861–1947), its first Reader establishes it against the odds; Andrew John Herbertson (1865–1915), holds the school together after Mackinder departs for a higher, national mission; and Kenneth Mason (1887–1976), the soldier and mountaineer, steers the school through difficult times during and after the Second World War, and directs its important war work (Coones 1989; Gilbert 1972; Scargill 1976; Goudie 1998 and Heyes 1887 for Mackinder, Herbertson, and Mason see *Geographers* 9, 3 and 18 respectively). In wider histories Oxford geography appears an adjunct to an empire of men: its most important early benefactions from the imperially minded and male Royal Geographical Society (RGS) (1899–1913) and Sir Abe Bailey, South African financier and politician (1910–15) enabled it 'to promote the study of the Geography and History of the British Empire'; its premises were a meeting point for male colonial officials on furlough; and the maps from the War Office, India Office and various colonial governments were 'of special importance' in its resources (Firth 1918 (quotation p.18); Scargill 1976, 455; Louis, Gadd and Eliot 2013, 370; Symonds 1986; Symonds 2000). Such histories record the establishment of its BA course (the Final Honour School or FHS, established in 1932 and first examined in 1933) as the mark of its firm arrival in Oxford (Firth 1918; Scargill 1976 and 1999), something that belittles its first qualifications (the diploma and certificate), although they suited many women better, being cheap, short courses, designed for school teachers, and for which there was no prerequisite in Greek which girls' schools rarely taught (Howarth 2000b; [Association for Promoting the Education of Women in Oxford] (1895)'). Such histories judge the FHS's success by the careers of its male graduates in 'the Colonial Office … the Church, the Army, and the Air Force … business and banking': teaching, the only profession of those open to women which the historian names, took last place on his list (Mason 1938, 412).

This then is the canonical history of geography at Oxford (Powell 2015a; Livingstone 1992), male when successful, female when not, with 'success' being calibrated by things that suited men; but there is another history, the history of women at Oxford, which this article seeks to tell (see Maddrell 2015 for women's different experience of geography's past). In some ways this is a 'hidden history' (Driver 2013), but one hidden in plain sight: for example, women appear in Firth's account of the effect of war on the School of Geography, but only as the total number of students minus the (regrettably) small number of men (1918, 22). In some ways this is a 'small story' (Lorimer 2003, and see Sheppard 1994; Ellis 2013; Prince 2000). It is the story of a subject considered marginal for much of its history (perennially short of money, for example (Firth 1918)) and of people occupying marginal positions within it: students on short courses (see table of diploma and certificate students), schoolmistresses, authors of geographical textbooks (**Fanny Herbertson** and **Dorothy Forsaith**), a librarian (**Elspeth Buxton**), teachers doing short-term work in summer schools (**Lettice MacMunn**) and an unofficial staff helper (**Fanny Herbertson**). But it is not only a small story, since from an early date women geographers were major figures in the Oxford department: this chapter enables a highly distinguished University Reader (**Marjorie Sweeting**) to take her place alongside the Reader around whom histories of the Oxford discipline typically pivot (Mackinder) and draws attention to the fact that she produced research of international quality in a department that was otherwise very poorly regarded by British academic geographers at the time (Halsey 1994, table 22.5 for 1976 figures). We describe a prolific geographical author who helped consolidate Oxford University Press's role as a leading educational publisher (**Fanny Herbertson**). We describe geographers distinguished as undergraduate students (**Dorothy Doveton**) and as doctoral students (**Marie Bentivoglio**). We describe one of the longest serving staff members in the school's history (**Nora MacMunn**) whose 28 years of service (1907–1935) outdo the combined service of the famous male duo of Mackinder (8 years, 1887–1905) and A. J. Herbertson (16 years, 1899–1915).

This essay is written with the benefit of the new perspective provided by the seminal *History of the University of Oxford*, which has revealed a university that was, in addition to the more obvious things, a trainer of unmatriculated students in vocational subjects on short courses; a vehicle for extension lectures to audiences outside Oxford; a major player in publishing, school examining and raising the standard of school teaching; and a source of intelligence in wartime (Brock and Curthoys 1997 and 2000; Harrison 1994 and see Goldman 1995; Ockwell and Pollins 2000; Mackinder and Sadler 1891). With the benefit of such a broad perspective we can produce a more rounded history of the school as a whole, and not simply a history that includes women in a narrative shaped by men's priorities. The new perspective can, for example, reposition A. J. Herbertson's turn from geographical research to geographical education less as an unalloyed personal 'sacrifice' (Gilbert 1972, 194), but, rather, as an exchange of one of the school's and the university's core activities for another (Jay 1965b, 1979).

Women at the University of Oxford

Women's admission to the University of Oxford has been the subject of considerable study (Howarth 2000b, on which this account is based unless otherwise attributed) and we summarize here only those trends that have a particular bearing on the study of geography at the university. The arrival of women at Oxford from the 1870s and 1880s was managed gradually to avoid competition with men (originally women did not sit for University examinations but took those of the Oxford Local Examinations Delegacy, for example), and to be at least partially separate from them (they lived at home or in separate halls and had their own lectures, though they could attend University lectures and intercollegiate tutorials if chaperoned and with permission of the lecturer in question).

Amid controversy about whether, where, with whom and how women were to be educated, there was considerable agreement in Oxford, including in conservative and church circles, on the need to educate women to be schoolmistresses, though many doubtless wished that such schoolmistresses might be trained outside Oxford. With secondary school teaching the principal career of 38 per cent of women at the Oxford halls before 1914, the School of Geography's early welcome to women already, or intending to become, schoolmistresses fits the wider Oxford, and, indeed, national, picture.

Various milestones on the journey to women's equality with men at Oxford and an end to sex discrimination can be identified: the opening of undergraduate examinations to women (from 1884); the establishment of women's halls (from 1879) and their development to resemble men's colleges, with collegiate buildings and college tutors (from the 1880s); the establishment of a Delegacy for Women Students, which brought women under official University supervision (1910); the allowing of women to matriculate as members of the University and study for almost all degrees (1920); the incorporation as colleges of the four residential societies for women (1925–1926) and their recognition as full 'colleges of the university' (1959); and the trend towards co-educational colleges (from 1974). Each step in this progress was hard-won and there were setbacks, including the imposition in 1927 of a quota on women (*Oxford University Gazette* 58, 1850 (1927), 112); it was raised in 1948 before being abandoned in 1957 (Howarth 1994, 350, 356–8).

A brief history of the University of Oxford's School of Geography, with particular respect to women

Women geographers at Oxford may have thus far escaped particular notice not simply for the reasons that women's history remains almost everywhere obscured: 'Let us now praise famous men,' exclaimed O. J. R. Howarth rapturously as he told the history of geography within the British Association for the Advancement of Science (1951, 145). Rather, women geographers at Oxford have also escaped remark because their admission was driven by expedience not ideology, so, as with their admission to the law faculty at Oxford (Auchmuty 2008), they were present from the start. Thus Mackinder, the anti-suffragist, whose intellectual stage was imperial, whose public stage was parliament, and who was at home among the RGS's male explorers, army and naval officers, and imperial servants, thought the high point of his first Oxford years the 'two or three men, with a strong inclination to Geography, who would make it a speciality were there sufficient opportunity' (letter of Mackinder, cited in Scargill 1976, 445; Baigent 2004b; Bell and McEwan 1996; Maddrell 2009, 30–35). However, in the absence of such an opportunity, and to keep numbers up, he waived the fees for actual and potential schoolmistresses attending his lectures (Scargill 1976, 445).[1] Similarly, A. J. Herbertson dreamt of a university geography school dedicated wholly to 'training the men who are to shape the destinies of their country and of their time' (1902, 124); but in the meantime he welcomed actual and potential schoolmistresses to Oxford's School of Geography (Carter and McGregor 1901–2). Seen from the heroic point of view, the school's history is an unsatisfying story of compromise, uncertainty and making do (Firth 1918); but even the least enlightened leader saw that women students were better than no students at all (actually only just, in the view of the school's first historian (Firth 1918, 22)). In these circumstances women from the start took advantage of the school's relative informality and openness to further their geographical education.

The establishment of geography at the University of Oxford followed a pattern typical of new subjects at Oxford (and geography departments around Britain) (for general histories see Firth 1918; Scargill 1976; and Coones 1989). Amid discussions about the broadening of the curriculum that had been going on at Oxford since the second half of the nineteenth century, individuals were appointed to university posts in new areas, but the establishment of a bachelor's degree (FHS), and with it coveted college fellowships and a sense among undergraduates that the disciplines were worth focusing on, had to wait (see, for example, Firth 1918, 22 on Herbertson's (unsuccessful) efforts to get a college fellowship at least for the professor of geography: it was not until 1967 that all tenured staff in the school had college fellowships (Roche 1994, 281)). Thus in 1887 Halford Mackinder was appointed Reader in Geography – a university not a college post and one created only with the financial backing of the RGS (Kearns 1985). Geography was at first a subsidiary subject in the FHSs of Lit. Hum. (classics) and modern history. Modern historians made up the bulk of Mackinder's students, and that provided a way in to geography for women, since modern history at Oxford was a favourite with women

(Firth 1918, 5; Howarth 2000b, 281–2). Mackinder lectured to 337 members of the university (i.e. men) between 1890 and 1895, and 77 'registered lady students' between 1892 and 1895 (cited in Firth 1918, 7). The figures are not directly comparable, but suggest that, from the start, nearly a quarter of those seeking geographical education (whether for geographical or other qualifications) were women.

Discussions in the RGS and Oxford bore fruit in the establishment of the Oxford School of Geography (1899, with Mackinder as Director), the Diploma in Geography (first examination 1901) and the certificate for those showing proficiency in a particular subject of the diploma course, such as surveying, regional geography or historical geography (1903) (Firth 1918, 10; Honey and Curthoys 2000, 564). The diploma itself comprised at first a compulsory paper in each of physical and regional geography, and three optional subjects from a list that included papers in geography (geomorphology), those designed for classicists (ancient historical geography) or historians (medieval and modern geography), and technical papers (surveying and cartography) (*The Geographical Teacher* 1,2 (1902), 90; Murray 2000). The diploma syllabus changed over the years, but a compulsory description of a selected district based on 'personal observation and the study of maps, statistical returns, books, etc.' (introduced as part of the regional geography option in 1906 (*Oxford University Gazette* 1905–1906, 410) and made compulsory in 1910 (University of Oxford Examination Statutes 1910, 247; 1916, 58–62) became a distinctive and enduring feature. (See Table 2.1 for statistics on women in the diploma and certificate courses.) Teaching took place in the lecture hall and classroom, and also in the field.

The records speak of the establishment of the diploma as a male project (for 'Honours men' (Firth 1918, 9)), and some courses were, indeed, designed primarily for men (surveying, though, in fact, women did surveying, too (Carter and McGregor 1901–1902)) or for those preparing for examinations open only to men (civil or military service) (Firth 1918, quotation at p.17; *The Geographical Teacher* 1,2 1902, 90). However, like other Oxford diplomas, the geography diploma attracted women from the start since diploma students did not have to matriculate.[2] Mackinder accordingly sent his diploma prospectus to girls' as well as boys' schools: 'My Committee feel that a man/lady possessing such a diploma ... should be a valuable accession to the staff of a school' (cited in Scargill 1976, 450). A. J. Herbertson stressed in his periodical *The Geographical Teacher*: 'Both men and women are admissible' at the Oxford school (1,2 (1902), 90). It is hard to establish whether the welcome offered by Mackinder and A. J. Herbertson to women was merely expedient. Certainly women were needed to make the course look flourishing and their fees were needed to keep the school solvent (though the geographical scholarship seems to have been reserved for men – and, indeed, for 'gentlemen' (Scargill 1976, 445). However, Mackinder and Herbertson had in 1893 helped found the Geographical Association (GA), an organization for school teachers, which admitted women from the start, and Herbertson nominated women for RGS fellowship (Bell and McEwan 1996), so there is some evidence of a genuine wish to include women in the geographical project. Whether from expedience, genuine openness or a mix of both, plans to improve geographical training at Oxford for schoolmasters undoubtedly advanced such training for schoolmistresses.

Women were also a notable presence on the biennial Oxford summer schools in geography held in the first three weeks in August, initiated by Mackinder in 1902 and continued by Herbertson, to enable schoolmasters and mistresses to improve their theoretical and practical knowledge of the discipline (Carter and McGregor 1901–1902; see Figures 2.1 and 2.2). The summer schools attracted speakers of note, women such as Ellen Semple (Keighren 2010, 97–100) among them, while Mary Croom Brown, historian and author of *Mary Tudor, Queen Of France* (1911), guided the students through the school's geographical literature – 'a capital feature of the course, one appreciated by all of us' (Carter and McGregor 1901–1902, 176). Field work was also much appreciated; field skills such as surveying were taught to women and men, and women and men shared equally in the camaraderie of field trips (Gilbert 1972, 193; *The Geographical Teacher* 1,3 (1902), 124; Carter and McGregor 1901–1902, 178). The first summer school attracted 30 students, but by 1916, 125 attended – a remarkable number considering the backdrop of war, which severely diminished the supply of schoolmasters (Firth 1918, 14, 22). More than 850 teachers attended the five schools organized by the Herbertsons between 1909 and 1914 and the influence they thus exerted on British school geography was immense (Gilbert 1972, 193). The establishment of these early geographical initiatives linked to improving

Figure 2.1 Geography teachers on a vacation course field trip, 1910. Credit: University of Oxford Archives.

Figure 2.2 Geography teachers on the 1922 vacation course organized by the School of Geography. Credit: University of Oxford Archives.

educational standards shows Mackinder and Herbertson among the Oxonians (not always in the majority) who wanted the University to advance the radical changes under way in education in England and Wales (Honey and Curthoys 2000). Their inclusion of women was entailed in this commitment to geographical education, which, in turn, was in part a by-product of the lack of opportunities for geographical research.

Geography benefitted from Oxford's introduction of the research degrees of BLitt and BSc in 1895 (Howarth 2000a, 618). By 1902 the school offered facilities for those preparing for the BSc examination (*Geographical Teacher* 1,2 1902, 90), and, by the time that Firth reported in 1918, three men had graduated BSc and one BLitt with work done substantially in the school (17). The degrees, requiring matriculation, were not open to women until 1920 (see Table 2.2).

By 1957 when the cap on women was abandoned, women formed only 13.7 per cent of the new undergraduate intake across the University as a whole (*Oxford University Gazette* 87(2911) (1956), 239–52). At least by 2004, when statistics on gender by course were published for the first time, more than half of the School of Geography's undergraduates were women, a situation reached across the University as a whole only in 2017 (2017 University of Oxford, Academic Administration Division, Student data management and analysis. 'Student statistics', https://www1.admin.ox.ac.uk/aad/studentregistry/sdma/statistics/student/; accessed 3 July 2018).

The fact that the geography FHS proved popular with women was largely a by-product of external factors: positive ones, such as the facts that woman could matriculate at Oxford by the time it was introduced and that school teaching offered women good employment prospects, and negative ones, such as the state of girls' education, which left them unable to read for some other degrees. Women's inclusion in the school was not a desideratum per se. Indeed, A. J. Herbertson argued for a geography that would be 'an honour school in natural science' (Currie 1994, 117), something which, like his 1902 proposals for a geography school training people for public life, would largely have excluded women, few of whom had the school training for science degrees (see Roche 1994 for women in the natural sciences in Oxford 1920–67). However, Herbertson's proposals attracted little support from Oxford's natural scientists, and instead, the new FHS syllabus largely followed that of the diploma with a regional emphasis and a stress on the 'geographical description'. The school grew very fast: by 1939 it was attracting 4 per cent of both female and male FHS students (Currie 1994, 117) and women undergraduates achieved early success (see Dorothy Doveton below).

Women staff members at the School of Geography

If women students had considerable success at Oxford, the position of women academics in the School of Geography was slower to improve – as it was in the University as a whole. Compared with the position of other salaried University staff, Nora MacMunn's 1906 appointment to the School of Geography was very early, but the school had to wait until 1951 before there was more than one woman academic member of staff (Marshall and Sweeting), until 1978 before there was a woman Reader (Sweeting), and until 2003–2004 before there were women professors (**Diana Margaret Liverman** (*b.*1954); **Linda Margaret McDowell** (*b.*1949) and **Sarah Jane Whatmore** (*b.*1959)). In spite of an overall increase in academic staff numbers, the ratio stubbornly remained between 18 and 22 per cent for women until the twenty-first century. Since the millennium, women have been appointed as Head of Department (Whatmore in 2012, **Heather Ann Viles** (*b.*1959) in 2015 and **Gillian Rose** (*b.*1962) in 2019) and by 2018 they formed 37 per cent of the salaried academic staff. However, what has emerged from the writing of this article is the amount of geographical teaching done by women who had no formal university appointment, particularly, but not exclusively, in the summer school. Women such as **Agnes Ramsay** taught a diploma and certificate subject (ancient historical geography) while **Fanny Herbertson, Madeleine Fripp** and **Lettice MacMunn** taught in the summer schools. Remarkably, at the 1910 summer school, of nineteen demonstrators, eleven were women and eight men, all recent students of the school ((*Oxford University Gazette* 1910–1911, 537–9). In addition, Fanny Herbertson was a prolific geographical author.[3] The present trend of patching together academic employments is foreshadowed here, and was something that affected the women's male colleagues, too, although to a lesser degree since almost all the more permanent posts went to men (Crawford 1955).

What follows is an attempt to make these women more visible. It comprises biographical memoirs of all the women who taught geography in the school and who died before 2015, listed in chronological order of date of birth. The length of each entry is proportional to the importance of each person in the life of the school and her eminence, but also to the existence of material already in print. Thus, Marjorie Sweeting, whose achievements deserve their own lengthy *GBS* memoir, is included at shorter length than one might expect because much has already been published about her, whereas Nora MacMunn has gone substantially unrecorded and so her life is recorded here in more detail.

Fanny Louisa Dorothea Herbertson (1864–1915)

Fanny Herbertson was born Fanny (or Frances) Louisa Dorothy (or Dorothea) Richardson at Leamington, Warwickshire, in the autumn of 1864, daughter of Job (John Collingwood) Richardson (1836–1889), journalist and newspaper editor, and Louisa Moore (Baigent 2004a from where all information comes unless otherwise attributed). She was educated at the King's High School, Warwick. In 1888 she graduated BA from the University of London with a first class degree in classics, being a prize winner and placed equal second in the university as a whole ('Presentation day at the University of London' 1889). Very large numbers of London graduates were external students, especially before London became a teaching university in the twentieth century, and Richardson was one of many whom Cheltenham Ladies' College, a leading girls' public school, prepared for the BA until the 1920s with great success (Janet Howarth, pers. comm.). Richardson continued at the college as classics mistress, rising two years later to be head of the department with a reputation for ability and strictness (Jay 1965a, 359). She met A. J. Herbertson at one of the famous summer schools run by the sociologist and city planner Patrick Geddes at the Outlook Tower in Edinburgh. The two married on 5 October 1893 at 3 North Fort Street, North Leith, according to the rites of the Evangelical Union church, after which she followed him in a practical sense, as he moved to take up academic posts at Manchester (1894–1896), Edinburgh (1896–1899) and Oxford (1899–1915), and in a disciplinary sense as she turned from classics and sociology to geography.

Following school reform in the 1902 Education Act and its successor Education Acts for England and Wales, new disciplines jostled for recognition in school and university curriculums (Louis 2013, 370, 446). University teachers were active in promoting their discipline in schools hoping thereby to ensure a ready supply of university students. Many geographers were among those who wrote for school children and their actual or future teachers, and led summer schools for teachers – activities that spoke of a keen strategic sense of the future of their discipline rather than a failure to establish a research culture in it. A. J. Herbertson was one such: an enthusiast for the biennial Oxford summer schools and a prolific author of school texts. Fanny was not simply his helper in these endeavours: she was an important player independently. A contemporary remembers the summer schools thus:

> Herbertson … the small active man, the eager encouraging teacher, restless, continually at work, never learning to play or rest; or Mrs Herbertson, busy, energetic – even more so than Herbertson, a driving force at the summer schools, always everywhere, keeping everyone up 'to scratch'. (Brown 1948, 107)

O. G. S. Crawford, sometime demonstrator in the school, remembers Fanny in the first decade of the twentieth century as a fellow staff member (1955, 62), not just an occasional helper, although she never held an official post. The Herbertsons were prolific authors of school textbooks, sometimes together, sometimes independently. Of her sole authored books, Fanny's seven-volume *Elementary Geography* (1909–1912) was identified as a 'particular success' among 'educational bestsellers' in the official history of the Oxford University Press (OUP, Louis 2013, 446). By 1972, 1.4 million copies of the Herbertsons' OUP textbooks had been issued (Gilbert 1972, 193), and the couple also published extensively with A & C Black of London. Black's 'Descriptive Geographies from Original Sources' series was strikingly innovative. Attempting 'to depict the world in the language of men who have seen it', Fanny (the author) selected and reproduced travellers' accounts, consular reports and geographers' descriptions. Andrew edited the works and provided (rather dull) introductions. Some of the 'men who have seen' the world were, in fact, women: Lady Brassey, for example, in *Central and South America with the West Indies* 1902). Fanny's choice of extracts was perspicacious: Darwin on the Galapagos, Wallace on Amazonia, Schomburgk on Guiana, Head on the pampas, Nordenskjöld on Tierra del Fuego and Markham on Peru, for example, are works of enduring value that would fit readily into any modern compilation.

Though eclipsed by the 4 million books clocked up by Lionel Lyde (see *Geographers* 30), an Oxford classicist then teaching geography at University College, London (Clout 2003, 249), the Herbertsons' books had a profound effect on geography, and their commercial success contributed to OUP's emergence as an educational publisher of note and, in turn, to the press's remittances to the university at large. Fanny wrote as 'F. D. Herbertson, B.A. Lond.' – not quite 'M.A. (Oxon.), Ph.D. (Freiburg i. B.)' as was Andrew – but still clearly authoritative. As such she was sometimes, inevitably, assumed to be a man: *Man and his Work* was reviewed by the *Guardian* as by 'Messrs. A. and F. Herbertson' (advertisement in front matter of *Central and South America with the West Indies* (1902). Fanny's independent authorship shows perhaps most notably in her posthumously published biography of the French sociologist Frédéric Le Play, which was reprinted in the decades after its initial publication and is still widely cited (1920; 1921; 1946; 1950). In this work her easy and rather erudite style shows more readily than in some of her textbooks, which can bristle with facts, and she was obviously attracted to Le Play whose ideas had so influenced Geddes and, in turn, had affected hers and her husband's, dedicated as they were to the idea of regional geography and the interactions of people and places. The 1950s reprints of her work on Le Play were edited by Geddes's follower Victor Branford, founder of the British Sociological Society, who evidently approved her presentation of Le Play. She also translated a study of the British labour question by another French sociologist Paul de Roussiers. In indifferent health for much of her life, she died of heart failure 'with startling suddenness' (Anon 1916, 211) at the family's country house at Spriggs Holly, Radnage, Buckinghamshire, on 15 August 1915 (Jay 1965a, 360) and was buried with her husband, who had died just a fortnight earlier, in St Cross churchyard, Oxford, beneath a stone which also memorializes their son Andrew Hunter (1894–1917) who died in action in the First World War (Beckit 1915; Fleure 1965; Gilbert 1965b and Mackinder, MacMunn and Elton 1916; Unstead 1965). They were survived by their daughter Margaret Alice Louise (1899–1993), who became a general practitioner in Liverpool ([School of Geography, Oxford] 2015).

Fanny Herbertson was a fairly formidable and forthright individual. As we have seen above, she was noted for the strictness of her school teaching and for keeping Oxford students on their toes, and an obituary recalled her as 'a remarkable person of considerable…strength of character' (Jay 1965a,

359). O. G. S. Crawford, who knew the Herbertsons very well, describes them as being 'not too happily' married (1955, 45), but certainly she was a good intellectual partner for her husband – one of the few in the tiny school – and their joint labours underpinned the school's success and, indeed, survival. Her 'considerable intellectual ability' was obvious to all (Jay 1965a, 359). As well as being a prize-winning classics graduate, she 'spoke German like a native' (Gilbert 1965a and b, 318), and translated works from French. She was an important exemplar of the influence of classics on geography at Oxford, shown also by the work of, for example, Henry Francis Pelham, a keen advocate of geography at the University, G. B. Grundy, who lectured in ancient history in the school from 1896 and was a cartographer, and Charles Lucas (1853–1931), editor of OUP's *Historical Geography of the British Colonies* (Louis 2013, 370; Schwartz and Przyblyski 2004, 146; Butlin 2004). As a successful classicist Fanny Herbertson fitted this mould well and made her mark on Oxford's geographical teaching and the shape of geographical school education (Ploszajska 1999), as well as holding a minor place in the history of British sociology.

Nora Eileen MacMunn (1875–1967)

Unlike Fanny Herbertson, Nora MacMunn (see Figure 2.3) did hold a University post, being only the second woman to be appointed to a teaching post by the University: the first also taught on a diploma course (Howarth 2000a, 612), a fact that underscores the conjunction of marginal people on marginal courses at this point in the University's development. MacMunn was additionally an activist for the cause of women's suffrage, and one of the few at Oxford who belonged to the militant branch of the suffrage movement.

Nora Eileen MacMunn was born on 25 June 1875 in Chelsea, the fifth child and third daughter of John Alexander MacMunn (1827–1899), M.D., army surgeon and Principal Medical Officer at Chelsea Hospital, and Charlotte Edith (1842–1927), daughter of Revd George Mathias (*d.*1884), sometime army officer and Chaplain to the Royal Hospital and latterly Chaplain to Queen Victoria (Anon 1899; Bromley and Bromley 2015). Their first child was George Fletcher, later Lieutenant-General Sir George Fletcher MacMunn (1869–1952). The children were brought up at the Royal Hospital (Creagh and Humphris 1924; G. F. MacMunn 1930) but the family moved

Figure 2.3 Students for the Diploma in Geography with their tutors (middle row) Nora MacMunn, H. O. Beckit and W. G. Kendrew, 1916. Credit: University of Oxford Archives.

to St Leonards, Sussex, in 1896 (3 St Paul's Place and then 1 Blomfield Road (Wojtczak 2002, n.p.)). MacMunn was educated privately and then at the University of Oxford, being a member of the Society of Home Students (*est.* 1879 to offer an Oxford education to those who could not afford the fees of other women's societies (*The Ship* 1935, 36)). While a student she was secretary of the 'Sharp Practice' society, an impromptu debating society where women had to speak with no preparation so they could gain the experience in oratory and political discourse denied them by the Oxford Union, which did not open to women until 1962 (*Fritillary*, Dec 1904, 527; Griffin 1986, fn to 45; Wiggins 2009; Howarth 1994, 355–6). MacMunn sat for the FHS examinations in modern history in 1903 (Maddrell 2009, 154–5), but as a woman was unable to graduate BA. She proceeded to the diploma in geography (Trinity Term 1904; *Fritillary*, Dec 1904, 527), again as a Home Student. In 1906 she was appointed Demonstrator in geography initially for two years, 1907–1909, but, in fact, she remained on the staff until she retired in 1935 (*Fritillary* 39, December 1906, 664; *Fritillary* 41, June 1907, 696). She was Senior Demonstrator by 1918 (Firth 1918), Assistant Lecturer in 1922 and Lecturer in regional geography in 1924 (Scargill 1976, 458). By 1909 she was giving lecture courses and, in 1933, supervised a research degree. Alongside her university posts she taught at the Society for Home Students. She served in 1912 on the Committee for the Society of Home Students, alongside the prominent Oxford activist for women's education and suffrage Annie Rogers (*Spectator* 15 June 1912, 949; *Suffrage Annual* 1913, 346). MacMunn graduated MA by decree (*Oxford University Gazette* 1902–21, 108); in order to graduate by the normal route women who had sat for FHS examinations also needed to sit for some other qualifying examinations; graduating by decree was a route opened to women principals and some women dons to prevent their having to do this (Janet Howarth, pers. comm.).

MacMunn lectured and published educational material in regional geography as was appropriate to university and school curriculums: her *Upper Thames Country and the Severn–Avon Plain* (1913) in the Oxford Geographies series, for example, was a well-written and illustrated text, which followed the series' descriptive format. An important work was the *Guide to Geographical Books and Appliances* (1909, with A. J. Herbertson and J. F. Unstead) for the Geographical Association. Maddrell notes that the book recommends Olive Schreiner's *Story of an African Farm* (1883) for further reading (2009, 157). This is a rare glimpse in print of MacMunn's lifelong radicalism. Schreiner was profoundly influential on the women's suffrage movement and in feminist politics more generally (see her *Woman and Labour* (1911), for example) and her politics lay close to MacMunn's heart. MacMunn was active in women's suffrage politics in both St Leonards (her parental home) and Oxford. Her mother Charlotte was honorary treasurer to the Hastings and St Leonards Women's Suffrage Propaganda League, while Nora and her sister Lettice also served on its committee. Charlotte handed out leaflets, Lettice and Nora marched in London for the cause, and in 1910 Nora read a historical paper before the league entitled 'The position of women, in relation to the development of the Constitution', which claimed that women's civic position had deteriorated (Kay 2007, 537; Wojtczak 2002, n.p.). Nora joined the Oxford University Women's Suffrage Society with whose members she marched in London in the University contingent of the big London rally in 1908 (*St Hugh's Chronicle* 8 (1935–36, 22). More remarkably, she was a member of the Oxford branch of the Women's

Social and Political Union (WSPU, 1911; Crawford 2006, 205; *Suffrage Annual* 1913, 298), the suffragettes who espoused direct action, including violent action, in support of their cause. *The Suffragette*, the WSPU's periodical, records her many donations to the militant suffrage cause (12 Nov 1912 10/-; 27 Dec 1912 5/-; 3 Jan 1913 5/-; 18 Apr 1913 £1 10/-; 12 Sept 1913 5/-; 20 March 1914 £5). Unlike the many women who withdrew support for the WSPU as violence increased, MacMunn persisted in, and indeed increased her financial support for it, though there is no evidence that she personally committed acts of violence. She was also a member of the Women's Tax Resistance League (WTRL, *fl.* 1909–1914; Crawford 2006, 205; *Suffrage Annual* 1913, 298). An offshoot of the Women's Freedom League (*est.* 1907), the WTRL invoked the long-standing call of 'no taxation without representation' to argue that taxing disenfranchised women was unconstitutional and tyrannous, and to link tax resistance to established themes of British liberty going back to Magna Carta. It thus followed its parent by advocating civil disobedience and passive resistance, but not violence to persons or damage to property (Frances 1998). Membership of both organizations marks MacMunn out as one of a minority of Oxford women prepared to support the more controversial suffragette, and not just the less controversial suffragist, movements. Non-payment of tax was particularly controversial for unmarried women like MacMunn: such women were liable to pay their own tax (married men were liable for their wives' tax, so wives could not break the law by withholding tax) (Crawford 2006, 205; *Suffrage Annual* 1913, 298; Liddington and Crawford 2014 and see Park 1988). We have not yet been able to discover if MacMunn did refuse to pay tax on her income from the School of Geography.

Joining the WSPU made MacMunn the only militant suffragette named in Howarth's groundbreaking history of women at Oxford (Howarth 2000b, 304). Katherine Bradley identifies a further staff member of the School of Geography who was a member of the Oxford branch of the WSPU (*est.* 1908): **Madeline Catherine Fripp (1877–1959)**, who was WSPU Literature Secretary and briefly Hon. Sec. (Bradley 1997, 222–3; Crawford 2006, 100). Fripp, who had taken Oxford's history FHS examinations from the Society of Home Students, by 1902, being placed in the third class (Anon 1902, 256), gained the geography certificate in 1909 and then registered for the diploma but appears not to have taken it, though she attended seminars in 1910 and 1911. She published in *The Geographical Teacher* in 1910, and was one of the first women fellows of the RGS (1913, proposed by A. J. Herbertson) (Bell and McEwan 1996, 301, 303, 309, who suggest that Fripp was American but this conflicts with other information). Fripp is absent from the 1911 census, which suggests that she was perhaps one of the suffrage activists who, led by the Women's Freedom League and to some extent by the WSPU, refused to complete their census forms and, instead, chose to *Vanish for the Vote* (Liddington, Crawford and Maund 2011; Liddington with Crawford 2014; their gazetteer does not record Fripp as a census boycotter but they acknowledge the difficulty of investigating a phenomenon that depends on negative evidence; Nora MacMunn was included in the census). Laura Mayhall's *Militant Suffrage Movement* (2003) places the boycott squarely within the radical tradition of seeing the unenfranchised as having the right to rebel against the state, something that makes clear quite how subversive the boycott was regarded at the time, notwithstanding its having been largely eclipsed in the established narrative of suffragism by more dramatic and violent protests. Fripp taught in the 1910 summer school (*Oxford University Gazette* 1910–1911, 537–9) and possibly on other occasions in the school (Bradley 1997, 222). In that same summer school, **Lettice MacMunn** was a demonstrator in field sketching (*Oxford University Gazette* 1910–1911, 537–9). (She went on to be Principal of the Teachers' Training Department of the Royal Drawing Society (1920) and Director of the Queen Anne's Studio, Chelsea (1923), and was an accomplished art teacher (Teachers' Registration Council (1914–1948), Registers. Mss Archives, Society of Genealogists, London)). Like Nora MacMunn, Lettice MacMunn supported the WSPU financially (*The Suffragette* 1 Aug 1913 £3, 5 Sep 1913 £3, and 19 Sept 1913 £3), as did Fripp (*Votes for Women* 26 April 1912 £1, *The Suffragette* 21 Feb 1913 2/6). The 7th annual report of WSPU (1913, TNA Kew, Home Office papers 45/10700/236973) shows that between March 1912 and February 1913 Lettice MacMunn gave £2, Nora MacMunn £4 15/- and Fripp £1 15/-.

With Fripp and the two MacMunn sisters, there were then three active campaigners for women's suffrage on the school's staff, a number exceeded in the University by only the Society for Home Students, which had five WSPU members (Fripp and Nora MacMunn were attached to both the society and the school). MacMunn was no stranger to risk-taking, having previously taken part in a street protest after which she had had to dodge the police to avoid arrest (Wojtczak 2002, n.p.). MacMunn's activism originally owed more to her family background and St Leonard's than to Oxford (Lettice had been active in the cause since 1904, three years before Nora joined in 1907 (*Suffrage Annual* 1913, 297–8)), and this fits the pattern noted by Howarth that it was not axiomatic that suffrage activism and support for promoting women's education in Oxford would go hand in hand (Howarth 2000b, 244–5, 303–5; Liddington and Crawford 2014, 192). It remains unknown whether and how far the women's views were known to their male colleagues in the school. Halford Mackinder, a vocal anti-suffragist (Harrison 1978, 76, 104; Kearns 2010, 193) had left Oxford by the time Nora MacMunn was appointed, and happily her junior in 1910, O. G. S. Crawford, declared himself 'a supporter of the suffragettes' (1955, 70). This was just as well, since one of his students was Dorothy Wharton (1880–1928), six years his senior, an experienced schoolmistress, and a suffragette. In 1912 was sentenced to a month's imprisonment for malicious damage, something which suggests that her absence from the 1911 census was intentional. A Home Student, she took the geography diploma with distinction in 1913 (1911 census; Teachers Registration Council, 1914–1948; Home Office, Registered Papers, Suffragettes Meetings, Outrages, etc.).

Maddrell (2009, 155) suggests that MacMunn was overtaken in academic status by her male colleagues: It is true that Henry Oliver Beckit was made acting director of the school in 1915 following Herbertson's sudden death, Reader in 1919 when the post was advertised after the war, and later head of the department (Scargill 1976, 458), while James Cossar (MA (Edinburgh), FRGS), came to Oxford in 1919 as assistant to the Reader, and ended his Oxford career as lecturer, with a spell as director of the school (Scargill 1976, 459; Steel 1987, 60). On the other hand, MacMunn was senior in rank to the first two junior demonstrators, both men: O. G. S. Crawford (1910–1912) and A. G. Ogilvie (1912–1914) (Scargill 1976, 455) and MacMunn's salary, which started at a scant £15 for being demonstrator in 1906, with a further £20 for bibliographic assistance, rose steadily to reach £60 in 1916 when she was the second highest salaried staff member in the school (Geography archives GE3, University of Oxford Archives). On the whole, however, she did not progress as far as long service and obvious intelligence suggest would have been appropriate, and the problem was not lack of opportunity since her senior male colleagues all died distressingly young (Herbertson at 50 in 1915, Beckit at 56 in 1931, and Cossar at 50 also in 1931), so opportunities for promotion opened fairly regularly.

MacMunn's experience was typical of Oxford where integration into the life of the university took longer for women dons than for women undergraduates.

Women dons, though comparatively numerous, were rarely asked to assume positions of responsibility, for example, to examine, and even by the late 1950s only one of Oxford's 97 professors was a woman (Howarth 1994, 353). MacMunn's failure to achieve a senior role in the school was not due to her having been trained as a historian – so were many other early Oxford geographers, for example, C. F. W. R. Gullick, Edmund William Gilbert and John Norman Leonard Baker (Steel 1987, 59–60) – and it was scarcely a disadvantage when most undergraduates taught in the school were modern historians, and the Regius Professor of Modern History, Charles Firth, was a staunch advocate of geography in Oxford and beyond (Firth 1918; Roots 2004). MacMunn was less well qualified than some of her male colleagues, largely, of course, because Oxford's degrees and the RGS's fellowship were closed to her until 1920 and 1913, respectively.

MacMunn continued to write as a historian as well as a geographer, and in so doing had her share of patronizing reviews. Her chapter on 'Social and economic history' for the authoritative *Victoria County History of Essex* (1907) (known to be by a woman as she used her full first name) was described as being 'lighter' than other chapters and containing 'many facts…which might not have appeared in more serious historical sections' (*Essex Review* 16, 1907, 104–5): in fact, it was densely packed with statistics and other information supported by exhaustive primary and secondary sources, in expected Victoria County History style. True to her social class and family background, her writing could not be caricatured as 'feminine', but, instead, shows her ready grasp of military history and strategy (her army officer brother was a noted author on India). On Gibraltar and Malta for the *Oxford Survey of Empire* (vol. 1 1914), for example, she writes:

> The immediate strategic importance of Malta has, however, waned slightly since the beginning of the twentieth century. The balance of political forces had changed, and with the rise of the German navy Britain has been led to concentrate her naval strength once more in home waters and reduce her Mediterranean fleet, depending on Gibraltar if need be to close that sea. Since the growth of continental naval power in the Inland Sea, the short sea route has ceased to be considered a war-time route to India. The absolute safety of the Cape route – for Gibraltar in our hands obviates the danger of any attack on our fleet from the Mediterranean – more than compensates for the extra fortnight of time required to reach India. (527)

Similarly, her work on 'The French and German borderlands, II: Geographical aspect of the political history' (1916) brought an authoritative view on the strategic significance of geopolitical history to a wartime audience. At home equally in history as in geography, she was commissioned by a joint committee of the Geographical and Historical Associations to review 'Historical maps and atlases', her resultant opinion being published in *The Geography Teacher* and as the Historical Association's pamphlet 13 in 1909. Though brief, the work shows her knowledge of the French and German literature.

While at Oxford, MacMunn made her home at 5 Bath Place, Holywell (*Suffrage Annual* 1913, 298). She retired in 1935 from her posts in the Society for Home Students and in the School of Geography and moved out of Oxford to Horse Shoe Cottage, Biddesden, near Andover, Hampshire (*The Ship* 1936, 16). The college record of her departure shows sadness going far beyond the formulaic expressions of regret required on such occasions: MDRL speaking for the society noted that 'all those who knew the part she played in the Society as a whole must regret our loss of her kindness…and thoughtfulness…We all of us owe her a great deal, possibly more even than we realized' (*The Ship* 25 1935, 29–30), while her pupils spoke of her as their 'beloved tutor and friend…Never did she fail us. We knew we could always turn to her in joy or in sorrow, certain that she would help us with her sympathy and wisdom…Her high standard of conduct in life meant far more to her students and friends than she could ever know' (*The Ship* 1935, 36–37). A tribute to MacMunn herself, these expressions also reveal the close, almost family, bonds of the early women's societies. At some point MacMunn shared a country house near Ludgershall, Buckinghamshire, with Ivy Williams (1877–1966), first woman barrister in England, who had studied and latterly taught at the Oxford Society for Home Students (Fox 2004). MacMunn died unmarried on 14 May 1967 at Eastbourne, East Sussex (*The London Gazette* 44336 (6 June 1967), p. 6414). She was not memorialized in the geographical press, being by that time long forgotten. She was, however, long the only salaried woman among a large cast of men who taught in the early school (including many occasional lecturers from other disciplines) (Firth 1918); but even her long service somehow counted against her: she seems merely left behind in the First World War while her male colleagues 'take up Government work of importance' or go on 'Active Service' (Firth 1918, 23). Her view of geography was firmly rooted in history, shaped by her degree, and informed by an understanding of military strategy, shaped by her family; but she nonetheless was, and regarded herself as, a professional geographer, being a member of the Institute of British Geographers (along with Dorothy Doveton) by 1936, just three years after it was formed as the academic counterweight to the RGS, which retained its focus on exploration and military service (*TIBG* 1936, ix). MacMunn's suffrage activism and that of her sister, Wharton and Fripp are counterweights to the presentation of Oxford geography as imperialist and conservative, and as long labouring under the *Legacy of Halford Mackinder* (Kearns 2009).

Agnes Margaret Ramsay (1875–1959)

Agnes Margaret Ramsay was born in 1875 in Smyrna, Turkey, to William (later Sir William) Mitchell Ramsay, classical archaeologist and later professor at Oxford and Aberdeen, and Agnes Marshall (*b.* 1850) (Anon 1939). Ramsay graduated MA from the University of Aberdeen in 1904 with first class honours in classics, winning the Dr Black Latin prize and the Seafield gold medal for Latin (Anon 1904). She then read literae humaniores (classics, 'greats') at Somerville College, attaining first class marks in her final examinations in 1907. She held academic posts in classics at the University of Aberdeen, Royal Holloway College, London, 1910–1914, and Bedford College for Women, London, 1914–18 (Anon 1914) and during the First World War served with the Army Education Scheme (Denecke 1959), before being appointed tutor in classics in 1921 at Lady Margaret Hall, Oxford. That year she lectured on the historical geography of Greece and Asia Minor at the School of Geography. Ancient historical geography was an optional paper for the geography diploma but also epitomized the role of geography as an ancillary subject for classicists. Ramsay had only a college (rather than a University or joint) post, but at Oxford such posts were and remain important in providing geographical teaching. Ramsay resigned from her post at LMH in 1931 for health reasons. She died unmarried on 14 July 1959, leaving an adopted son.

Mary Marshall (1917–1983)

Mary Marshall (Figure 2.4) was born on 5 April 1917, only daughter of Walter Marshall and his wife. She was educated at the Municipal High School,

Figure 2.4 Mary Marshall photographed in 1963. Credit: E.W. Gilbert's personal files, by kind permission of the School of Geography and the Environment Alumni Office.

Doncaster, and in 1936 she matriculated from Lady Margaret Hall (LMH) as an open exhibitioner to read for the BA in geography (Buxton 1984 on which all information is based unless otherwise attributed). Her father was a machine tool fitter at the time of her matriculation (matriculation form, School of Geography archive). She graduated with a first class degree in 1939, was awarded the H. O. Beckit Memorial Prize, and was appointed research scholar at LMH and to LMH's Maude Royden travelling scholarship. She began research on the Canal du Midi in southern France (unpublished typescript in School of Geography library in 1984) but was forced to return home when war broke out. Resigning her research scholarship, she served as Cartographer to the Oxford War Agricultural Committee before being appointed, in 1941, Geographer to the Naval Intelligence Division of the Admiralty, which was producing Naval Intelligence Handbooks. Clout and Gosme (2003) describe the Naval Intelligence Handbook project as 'a monument in geographical writing…the largest single body of geographical writing ever published'. The object was to provide detailed accounts of areas to where service personnel might find themselves posted: pithy and detailed, comprehensive and accurate, they took descriptive geographical writing to new heights. Space in the School of Geography was given over to the work, which saw large numbers of academics, from the school and beyond, as well as others with particular regional expertise, produce 28 volumes. Marshall contributed to the volumes on *Turkey* (1942–1943, 2 vols) and *Western Arabia and the Red Sea* (1946). Buxton (1984) suggests that she also contributed to *Palestine and Transjordan* (1943), and it is quite likely that, in the fevered wartime atmosphere, many authors contributed in one way or another to volumes that did not bear their names. Several of those who worked on the handbooks went on after the war to develop their work in the areas that they had covered (Clout and Gosme 2003, 166–7), but Marshall was not among them and her attention to those areas seems to have been due to wartime exigency rather than inherent interest or previous expertise.

At the same time as working on the handbooks, Marshall completed her Land Utilisation Survey report on Oxfordshire (published 1943). The survey, the project of L. Dudley Stamp (see *Geographers* 12), created a detailed inventory of British land and its uses in the 1930s and 1940s (Stamp 1937). In the same year she was appointed examiner in the shortened wartime Honour School, an important signifier of the competence of, and confidence in, a staff member – though, of course, it being wartime, the department was short-handed. By night she did fire-watching (night time observation on roofs or up towers to alert the fire brigade if incendiary bombs were dropped). In 1944 she was appointed lecturer in geography at LMH and from 1948 the Susette Taylor Fellow, named after Susette Martha Taylor (1860–1920), an LMH alumna who, a modern linguist and traveller, was in the first cohort of women to be elected FRGS (Middleton 2004; Howarth 2000b, 294). Elspeth Buxton remembers Marshall as an 'exacting but extremely stimulating' tutorial tutor, and as kind and attentive to her past pupils – even to the extent of giving them financial help. In 1945 she was appointed Departmental Lecturer at the School of Geography, and promoted Senior Lecturer in 1954, which post she held until her early retirement in 1982. Her lectures were remembered by Buxton as being up to date but lacking the 'sparkle' of her tutorials and field trip talks (1984). By the 1970s her main teaching was in helping with aerial photography classes (S. Squibb, pers. comm.). She helped lead field trips to Montpellier (Maddrell 2009, 244) but suffered from poor health, notably tinnitus, which led to nocturnal working hours as she sought peace and quiet (Heather Viles, pers. comm.). Her lecturership at LMH ended in 1961, but she was elected to an honorary research fellowship there the same year (LMH Register). This seems to have been a sop offered in apology for having ended her college lecturership rather than in expectation that she would undertake research. In continuing poor health, she retired in 1982 and died on 7 December 1983.

Marshall's career epitomizes the School of Geography as a backwater, left behind while other British university departments embraced the new research culture and modern curriculums. Geoffrey Hawthorn, later a distinguished social and political theorist at Cambridge, made the rather acerbic comment that Oxford regional geographers of the 1960s were 'chasing a chimera and an extremely uninteresting one at that' (1965), and even the kindly John Roche described the changes wrought in the department's personnel and outlook by the metropolitan Frenchman Jean Gottmann (see *Geographers* 25) from the 1970s as 'long overdue modernization' (1994, 281). Regional descriptions remained the staple of the Oxford undergraduate curriculum even in the 1970s, with special permission needed to deviate from the norm (Susan Squibb, pers. comm.), and Marshall's lack of research and failure to embrace new trends in the discipline was a department-wide phenomenon for which she was only marginally responsible. Her career also epitomizes the dying, but in her lifetime still abundant, breed of college tutors who were appointed to college teaching posts on the basis of their good performance in the FHS with no particular expectation that they would pursue research interests. Marshall was no less productive than many of her male colleagues, though she differed from them in lacking a college tutorial fellowship – a crucial mark of success in Oxford. (In 1965, she was elected fellow of St Cross College (Andrew Goudie, pers. comm.) but this was not a tutorial fellowship and the college not an Oxford college in the traditional sense of the word: further evidence if any was needed of women's marginalization in Oxford.) Clout and Gosme suggest that many geographers on the Naval Handbooks project never recovered their health after the pressures of wartime writing (2003, 170 fn 22). Marshall may have been one such. Certainly the post-war School of Geography offered no opportunity for participation in the kind of large collaborative project with fixed objectives and a clear publication structure in which she had been so productive during the war. Seen by many undergraduates as a typical 'old school' tutor (and frankly one rather 'past it' (Anon, pers. comm.)), she might ironically have flourished in the modern research culture of group projects, collaborative work, and clear agendas and deadlines.

Marjorie Mary Sweeting (1920–1994)

Marjorie Mary Sweeting (see Figure 2.5) was born on 28 February 1920 in Fulham, London, only child of George Scotland Sweeting, geology lecturer at Imperial College, University of London, and his wife Ellen Louisa Liddiard (all information is taken from Goudie 1995, Goudie, Ford and Sugden 1987, Kennedy 1995 and Viles 1996 unless otherwise attributed). Marjorie was educated at Mayfield School, Putney, London, and then at Newnham College, University of Cambridge, from where she graduated BA in geography in 1941, with first class honours. She stayed at Newnham as a research student until 1942, before becoming a geography teacher at Howell's School, Denbigh, Wales, for the remainder of the war. She returned to Newnham as Marion Kennedy Research Student in 1945. Supervised by geomorphologist Alfred Steers (see *Geographers* 34), Sweeting graduated PhD from the University of Cambridge in 1948 with a thesis entitled 'The landforms of the carboniferous limestone of the Ingleborough District, N.W. Yorkshire'. Between 1948 and 1951 she was a research fellow at Newnham College, Cambridge. In 1951 Sweeting moved to Oxford to take up the post of Lecturer and Director of Studies in geography at St Hugh's College. She was promoted to University Lecturer in 1953 and

Figure 2.5 Marjorie Sweeting, photographed in St Hugh's College. Credit: © Deborah Elliot, by kind permission of the Principal and Fellows of St Hugh's College, Oxford.

Reader in 1977. In 1983 she served as acting head of the School of Geography following the retirement of Jean Gottmann, being the first woman to assume this role. During her career in the University of Oxford she also served on the General Board, as well as taking on many roles in St Hugh's College, such as Dean and Senior Tutor.

It was, however, as a research scientist that Sweeting made her name at Oxford. In a department characterized by internal appointments of people trained in the regional geographical tradition, Cambridge-trained Sweeting was a rarity and, in her uncompromising commitment to rigorous scientific research, a harbinger of change. Sweeting was described by Jane Hart (2007, 86–87) as 'probably the most well-known of British women geomorphologists in the twentieth century', but this depressingly gendered comment obscures that fact that she was simply one of the best-known geomorphologists of the century. Her renown rested on almost 40 years spent researching karst landforms and landscapes around the world, and her prolific publications: about 70 of them, including two important books: *Karst Landforms* (1972) and *The Karst of China* (1995). Sweeting's early research and publications focused on temperate karst, but in 1955 she spent two months in Jamaica and this marked the start of her life's work on tropical karst. Her tropical karst research can be divided into three main periods: a phase of exploration and reconnaissance (1950s and 1960s); a phase of process measurement (1960s to 1980s), and a Chinese phase (1980s until her death). In Jamaica she carried out an aneroid barometer transect of the cockpit country and elucidated the importance of geology to the karst landscape. In the Kimberley area of north-west Australia, she undertook further physically challenging fieldwork in the late 1950s with Joe Jennings, including making several cave surveys. In 1977–1978 Sweeting led the geomorphology team that formed part of the RGS's expedition to Mulu in Sarawak, Malaysia, during which they undertook many measurements of the rate of karst dissolution processes. In the 1980s Sweeting collaborated with the Harwell laboratory just south of Oxford in applying isotopic dating and tracing methods to answer questions about the genesis and functioning of karst landscapes. She first visited China in 1977 when western geoscientists were admitted for the first time since the communist revolution. By the time she retired in 1987 she had made numerous trips there, visited many Chinese karst areas, given lectures to large audiences in Beijing, Guilin,

Liuzhou, Guiyang, Nanjing and Nanchang, and, as Professor Yuan Daoxian put it, 'done her best to act as an academic bridge between Chinese and Western geographers, particularly karstologists' (Yuan Daoxian 1987, 448).

Sweeting was always keen to support, encourage and apply to karst research new ideas and techniques such as radiometric dating and stable isotopes. She was impressively well-connected, and well-regarded internationally. She was the first woman chair of the British Geomorphology Research Group (from 1973 to 1975). Sweeting did not propose any major new theoretical frameworks, and her approach to geomorphology is encapsulated well in her statement that 'large global schemes of landform development will not advance our knowledge, it is detailed studies set in the context of a wider field which will' (Sweeting 1976, 4). She was extremely diplomatic and often unwilling to advocate one theory or perspective over another. However, her research outputs have had long-lived significance to, and impact on, the field of karst geomorphology. Her landmark book *Karst Landforms* (1972b) continues to be cited over 40 years after publication, and several of her early papers had considerable impact in documenting the nature of, and controls on, karst in different parts of the world. Her contributions to the twentieth-century development of cave and karst studies are contextualized by Ford (2016) and White (2016).

Sweeting also made a great impact as a teacher on students at all levels as her enthusiasm for her subject was infectious (Lorna Johns, pers. comm.). During her brief spell as a school mistress she inspired her pupil Jean Grove, later a distinguished physical geographer based at the University of Cambridge, to study geography at university (Johnston 2005, 658). As a tutor of undergraduates at St Hugh's she inspired her pupils to aim high, regardless of whether their interests were hers. She always encouraged her women students to take on challenging dissertations or apply for the Henrietta Hutton travelling scholarship, which she helped to administer. Resourcefulness and enterprise were qualities she admired and fostered in others. She expected her students to make their own reading lists, for example, but this was not resented by her students who recognized her genuine interest in their well-being (Susan Squibb, Elizabeth Baigent, Lorna Johns, pers. comm.). Committed to her subject, she built up the teaching of geography in St Hugh's to ten undergraduates a year, making it one of the most important and successful colleges in the university for the study of geography, but she was an old-fashioned tutor who did not coach students for examination success (Elizabeth Baigent, pers. comm.). She was particularly keen on fieldtrips, running many Saturday morning trips in the 1950s and 1960s to interesting geomorphological sites near Oxford, as well as running residential trips to karst areas such as the Burren in Ireland and Malham in Yorkshire. Though always supportive of women, she favoured the admission of men to the college teaching staff (John Wilkinson joined her to supervise the teaching of human geography in the college) and to the student body. Just as she thought the Oxford regional geographical emphasis a thing of the past, she thought that women's colleges should embrace modern changes. High standards, not traditions, were the important thing for her (Elizabeth Baigent, pers. comm.). Finally, she supervised more than thirty graduate students and gave much-appreciated support to international visitors to Oxford. Many of her students and mentees went on to have successful research careers. Derek Ford and Paul Williams, for example, two extremely influential figures in international karst geomorphology, were supervised by Sweeting at Oxford and acknowledge her role as mentor, as well as her enthusiasm, scholarly example, scientific integrity and wise counsel in the preface to their book *Karst Hydrogeology and Geomorphology* (2007).

Special issues of *Earth Surface Processes* and *Landforms* in 1987 (Goudie, Ford and Sugden 1987) and *Zeitschrift für Geomorphologie* in 1997 marking her retirement and death, respectively, contained numerous tributes from former students and colleagues and testify to her influence on international geomorphology. Sweeting received many national and international honours including the RGS's Gill Memorial Prize (1955), the Certificate of Merit of the National Speleological Society of America (1959), Honorary Membership of the Cave Research Foundation of America (1969), Medallist of Palacky University, Czechoslovakia (1973), Honorary Fellow of the National Speleological Society of America (1976), and the RGS's Busk Medal (1980).

Sweeting never married and lived alone in college and in the village of Enstone, north of Oxford. She enjoyed sport, particular Formula 1 motor racing, opera and socializing. The fellowship of St Hugh's College where she had spent much of her career meant much to her. With dark, tightly curled hair and spectacles, she was a convivial member of the common room. Indeed, part of what made Sweeting such a successful teacher and mentor and so successful in fostering international collaborations in challenging political circumstances was her sociability. Her generous hospitality was evident, for example, in her annual party for geographers at St Hugh's ('Never let it be said that a Sweeting sherry party ran dry!' she once memorably exclaimed (Lorna Johns, pers. comm.) and early evening tutorials, when undergraduates were offered sherry, were memorable (Susan Squibb, Elizabeth Baigent, pers. comm.). In retirement she continued to visit college for high table lunch and dinner, loudly lamenting her doctors' insistence that she lose weight to allow hip replacement operations to restore some of the mobility that she lost to arthritis towards the end of her life, when a bent posture and the use of a stick became habitual. She battled not only with mobility but also with Mandarin, determined to be able to read the Chinese karst literature more effectively (Elizabeth Baigent, pers. comm.).

Sweeting died of cancer on 31 December 1994 at the John Radcliffe Hospital, Oxford. She was buried in Enstone, Oxfordshire, on 6 January 1995. A memorial event (not a memorial service since she had no evident religious beliefs) was organized in St Hugh's at which Peter Bull, geomorphologist from the School of Geography, gave the oration and the playing of the pipa (Chinese lute) recalled her pioneering contact with China (Elizabeth Baigent, pers. com). Her estate was divided equally among St Hugh's College, Oxford, Newnham College, Cambridge and the RGS to provide funds to help travel for research. The British Society for Geomorphology named its prize for the best undergraduate dissertation in geomorphology the Marjorie Sweeting Award in her honour.

Elizabeth Mary Buxton (1926–2009)

Elizabeth Mary Buxton (see Figure 2.6), known as Elspeth, was born in Oxford on 12 March 1926 to Leonard Halford Dudley Buxton, lecturer and later Reader in physical anthropology at the University of Oxford and bursar of Exeter College, and his wife Marie Louise, née Montgomery, and orphaned in 1939 when both her parents died within 48 hours of each other from influenza complicated by pneumonia (*Oxford Mail* 3192 (6 March 1939) p.1, col.4). She was educated at Berkhamsted School for Girls (matriculation form, School of Geography archives) and in 1944 matriculated from Lady Margaret Hall, to read for a degree in geography. In 1945 she was recommended to Thomas Sharp, the architect and town planner, as an assistant for his work on the

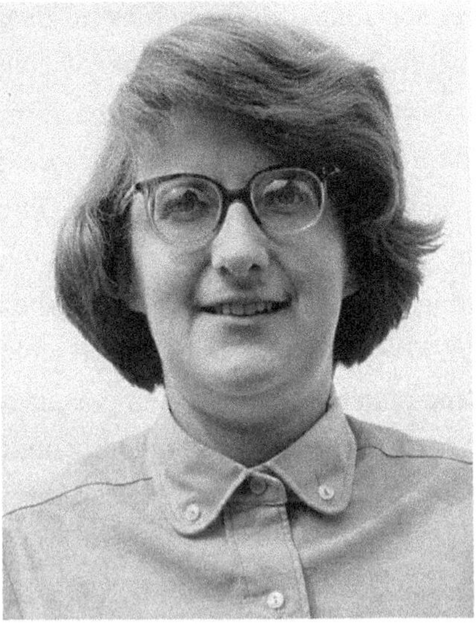

Figure 2.6 Elspeth Buxton, photographed in the 1960s. Credit: E.W. Gilbert's personal files, by kind permission of the School of Geography and the Environment Alumni Office.

replanning of Oxford (https://archiveshub.jisc.ac.uk/data/gb186-ths/ths5.4.12; consulted 8 July 2018). Although Sharp was happy in principle to employ her, the summer of 1945 saw him working on his Plan for Taunton (Newcastle, Thomas Sharp Papers (1945), 5.4.13 (MSS)) rather than on that for Oxford and her services were not required. She was appointed Assistant Librarian and Map Curator in the School of Geography in April 1953 and passed her Associateship examinations of the Library Association (part 1 1955, part 2 1956), and was registered as a Chartered Librarian in May 1957 (*Library Association Record* 57(9) (1955), 378; 58(3) (1956), 125; 58(9) (1956), 366–7; 59(5) (1957), 173). In Oxford colleges and departments of the time, the post of Librarian was customarily held by a member of the academic staff who was supposed to shape the academic direction of the library, while posts that entailed real library work were regularly advertised as being 'suitable for the wife of a senior member' with 'previous experience of library work desirable but not essential' (*Oxford University Gazette* 97 (3297 supplement) (1966), 377–8). Thus at the School of Geography the librarian to whom Buxton was notionally assistant was a (male) don: Freddie (A. F.) Martin 1940–1959 and Frank (F. V.) Emery 1960–1964. Buxton's professionalism was, however, acknowledged as she became a member of the sub-faculty of geography (that is, the body which oversaw the school's affairs) in spring 1961 and in 1964 she was promoted Librarian and Map Curator (with Emery as Senior Librarian in 1964), before finally gaining charge of the library in 1965 with no male don notionally her superior. In 1966 she was made Fellow of the Library Association by thesis on the subject of 'Government and other official publications for social and economic geography' (*Library Association Record* 68 (10) (1966), 368).

Buxton was described in 1945 as having 'the disadvantage of being six feet high' but as being 'a very clever, charming girl who is very keen on her subject and has the scholarly outlook' (Newcastle, Thomas Sharp Papers (1945), 5.4.12 (MSS)). She was always approachable and her willingness to share her deep knowledge of the extensive collections in the School of Geography's library is manifest in frequent grateful acknowledgements in books and theses. Robert Steel, for example, recalls 'the ready help forthcoming from Elspeth Buxton, librarian' (1987, xi). She was reluctant to discard material from the library's collections with the result that they became one of the largest specialist collections of geographic material in Europe, worthy of mention in the University's official history (Roche 1994, 280; Parry 2006). Her office was a further testament to her reluctance to part with material, to the consternation of successive fire officers at the school. She was a product of the pre-digital age, and generations of undergraduates recall her 'optical coincidence system' and its attendant 'knitting needles' with equal measures of awe and amusement, but the more perspicacious of them learnt to take advantage of her impressive knowledge of the geographical literature, something that was much valued in the era before computer catalogues and databases (Elizabeth Baigent, pers. comm.). With age she became extremely shortsighted and developed a pronounced hump on her upper back, but she continued to bicycle around Oxford, wearing a voluminous cycle cape in the rain (Lorna Johns, pers. comm.). Buxton retired in 1991 but continued to take an active interest in map librarianship and from time to time attended the Oxford Seminars in Cartography convened by Elizabeth Baigent of the School of Geography and Nick Millea of the Map Room at the Bodleian Library. She enjoyed gardening on the Trap Grounds allotments (Elizabeth Baigent, pers. comm.). Buxton died, unmarried, on 9 June 2009, in Oxford, having lived in a cottage in Hayfield Road, Jericho, Oxford, for over 30 years. Her near neighbours were **Sheila O'Clarey,** long-serving secretary to the head of the School of Geography, and Catherine Robinson, with whom she wrote a book about her street (1993).

A Roman Catholic requiem mass was held for Buxton at the Oxford Oratory Church of St Aloysius Gonzaga, Woodstock Road, Oxford (Parish priest's report, Oxford Oratory, 2009 www.oxfordoratory.org.uk/downloads/pprep ort09.pdf, consulted December 2018).

Mary E. Potter (*fl.* 1939–1987), artist, cartoonist and illustrator, was one of the 'Map Girls' identified by Maddrell in her 2008 study of work done by women geographers in the Second World War. Potter studied at the Slade School of Art, London, before returning to Oxford as cartographer to the Admiralty, being in charge of the Oxford drawing office for the Naval Intelligence Handbooks and 'responsible for many of the text figures and maps in our handbooks [at Oxford]' (Kenneth Mason, cited in Maddrell 2008, 134). Maps and illustrations were a required part of each handbook (Clout and Gosme 2003, 160). Potter did not achieve the later artistic fame of Gwen Raverat, who worked in the equivalent Cambridge office, but she remained more closely tied to the geographical world, working as a cartographer in the School of Geography. Within Oxford she was also a much-appreciated illustrator and cartoonist, notably for *Daily Information*, and producer of witty local maps for a general audience, as seen in her *Potter Pourri: 36 Sketches of Oxford Life* (1987). Her work, mostly black and white line drawing, was largely anonymous, but is immediately recognizable.

Barbara Anne Kennedy (1943–2014)

Barbara Anne Kennedy (see Figure 2.7) was born on 27 August 1943 in Ottawa, Canada, to an English father and a Canadian mother (Kennedy 1997 and *St Hugh's Chronicle* 2013–2014, 61–63, from where all information comes unless otherwise attributed). She had at least one sibling, David. Her father had been posted to Canada during the Second World War to train pilots, but the family returned to the UK when the war ended. She attended school in Worcester and Walthamstow School for girls, Sevenoaks, Kent. She read geography at Newnham College, Cambridge, from 1962 to 1965, being awarded a scholarship in 1964 and graduating with a double first in 1965. She returned to Canada as a Commonwealth Scholar (she was a dual citizen and proud of her Canadian links, Elizabeth Baigent, pers. comm.) to read for an MA at the University of British Columbia, Canada from 1965 to 1967, working under Mark Melton to produce a comparative

Figure 2.7 Barbara Kennedy, photographed in her office at St Hugh's College. Credit: © Deborah Elliot, by kind permission of the Principal and Fellows of St Hugh's College, Oxford.

study of multiply-controlled valley asymmetry in south-east Wyoming and south-west Manitoba. She then returned to the University of Cambridge to read for a PhD on valley-side asymmetry under the supervision of Richard Chorley, graduating in 1969. From 1968 to 1972 she was Lecturer and Fellow at New Hall, Cambridge, and simultaneously Demonstrator in the University of Cambridge's Department of Geography. She was Lecturer at the University of Manchester from 1973 to 1979 before, in 1979, being appointed CUF (Combined University Fund) Lecturer in Geography at the School of Geography and Fellow and Tutor at St Hugh's College. At St Hugh's she was Tutor for Admissions and then Senior Tutor (1996–2000) and won a reputation for her attention to detail in these roles and on a range of University committees. In 2000 she became the University's International Recruitment Coordinator, in which capacity she travelled extensively, but maintained connections with geography and St Hugh's. She retired in 2005, becoming emeritus fellow of the college.

Kennedy published sparingly (usually on her own) on topics relating to fluvial geomorphology and the history and philosophy of geomorphology, including two important books: one at the start of her career (*Physical Geography: A Systems Approach* with Richard Chorley, 1971) and one at the end (*Inventing the*

Earth: Ideas on Landscape Development since 1740 (2008). The first became a classic in geography, but a highly controversial one. It proposed a systems-based approach to the landscape. Mike Church (2014, 1270) described it as 'an intellectual explosion', but Kennedy quite quickly retreated from the manifesto it proposed and became a harsh critic of the abstract approach it championed. Indeed, her contribution to scholarship and geomorphological research was largely one of critique. She had an extremely incisive mind and was quick to spot flabby logic or fallacious argument. Her substantive geomorphological research focused largely on the structure of drainage networks and channel junctions. Her more important work, however, was a critical reappraisal of the history of geomorphological enquiry from the late-eighteenth and early-nineteenth centuries onwards, which culminated in her 2008 book. She was particularly interested in the work of Charles Darwin and its implications for geomorphological research. Her scholarly output, while slim, is provocative and rich in ideas. From the first to the second book, her research crossed from an explanatory framework that regarded geomorphology as process-based, and time-independent to one that focused on the historical, contingent nature of landscapes.

Kennedy was a demanding but inspiring teacher who instilled into several generations of geography undergraduates the importance of critical thinking. She could be terrifying in the face of indolence and poorly prepared essays ('The bluffness of her attitude to recalcitrant students was no bluff' in the words of her college obituarist (*St Hugh's Chronicle* 2013–2014, 62), but many of her students also recall her as being extremely hospitable and supportive (Church 2014). Being more at home with critique than data collection, she did not attract a cohort of graduate students, and was something of a lone scholar.

Kennedy was a marvellous cook, enjoyed good wine (she was wine steward of New Hall and St Hugh's) and opera, was fond of cats, and was extremely well travelled. She lived in college during term time, much valuing the companionship of her college where 'the Senior Common Room and the breakfast room were arenas for the force of Barbara's personality as much as was the governing body' (*St Hugh's Chronicle*, 2013–2014, 62). Outside term she lived in her houses in Chipping Norton, Oxfordshire, where she was governor of the town's secondary school, and in Langres, France, latterly with her partner, Adrianne. Kennedy died in Oxford on 4 February 2014 from cancer. Her funeral was held on 18 February 2014 in St Hugh's College chapel and a memorial service was held at the University Church of St Mary on 3 October 2014.

Notable Oxford women geography students

As one of the earliest established geography schools in Europe, the Oxford School of Geography has produced many notable alumnæ, some of whom are memorialized in *Geographers* in the current (**Doreen Massey**, **Helen Wallis**) or past (**Eva Germaine Rimington Taylor** vol.31) volumes. Others remain little known, for example **Blanche Hosgood** (*d.*1953), despite having gained the Oxford diploma in 1914 with distinction and having been head of the geography department at Bedford College, London, for more than thirty years. In the words of Dora Smee, Hosgood 'gave herself to Bedford College' (Smee 1953), a life of service, which was not the route to fame. Our intention here is neither to reproduce material on women already memorialized in *GBS* nor to attempt a complete record of women geographers of distinction who once attended Oxford, but, rather, to give short memoirs of some representative women students, listed in chronological order of date of birth.

Unlike the listing of staff, which is comprehensive, this section is selective and adopts the principle of 'representative biography'. For collective biographies that adopt the principle of 'noteworthiness', the *Oxford Dictionary of National Biography* for example, the expectation is that those included will be sought by name. In collective biographies that adopt in addition the principle of representativeness, those whose names previously are probably unknown are included to demonstrate some aspect of the life of the collective which is thought worth noticing. A notable collective biography that follows this principle is the *Australian Dictionary of Biography*, which includes, for example, rabbit catchers whose lives epitomized a moment in the life of the new country even though they, at least before they were memorialized, were not known by name (Arthur 2014; Arthur 2015). Although *GBS* generally adopts the 'noteworthiness' principle, it has had a foray into representative biography (the Hull House geographers in vol. 33), and this chapter takes the process further. The women listed here comprise such representatives: the sole woman in a photograph that has come to epitomize the early school (**Joan Reynolds**);

a writer of text books and trainer of teachers, representing the focus on school geography, which long shaped the school's priorities (**Dorothy Forsaith**); the second woman to graduate DPhil from Oxford – a degree that marked the university's entrance into the international research culture – but for whom it was Oxford's geography diploma that shaped her career (**Marie Bentivoglio**); the first person to gain a first class mark in the FHS examinations in geography, a qualification longed for for decades and which marked geography's coming of age in Oxford (**Dorothy Doveton**); and a scientist of international repute who trained at the school before going into a more clearly research-oriented field (**Moira Dunbar**). Length here is proportionate to the part the school played in the woman's life, the part the woman played in the life of the school, and whether the woman is memorialized at length elsewhere.

Joan Berenice Reynolds, 1870–1950

Joan Berenice Reynolds's (see Figure 2.8) biography is less important to geography at Oxford than is her iconography. She is the sole woman in a photograph of the 1901 Oxford geographical diploma class, a photograph that has come to epitomize the discipline's early establishment at Oxford. She sits surrounded by men, including the Great Men of Oxford geography: Mackinder and Herbertson.

Reynolds was born in Warwick in 1870, the third child of Frederick Reynolds (1833–1900), accountant and metal broker in West Bromwich, and Gulielma Metford, daughter of William Metford and Sarah Clark (1832–1889) (http://www.pennyghael.org.uk/Reynolds2.pdf;). She was educated at the King's High School in Warwick, as was Fanny Richardson, later Herbertson (Maddrell 2009, 129). She graduated BA from University College, Cardiff, University of Wales, and was assistant mistress at the Cardiff intermediate schools and winner of a Gilchrist Travelling Studentship for a visit to Switzerland and northern Italy in 1898 (Anon 1897, *South Wales Daily News* 23 October 1897) and published author before going to the Society for Home Students in Oxford to read for the diploma in geography. In Oxford she spent much time with the Herbertsons and, after she had left (to continue her teaching career), the three collaborated on *Descriptive Geography of Australia and Oceania from Original Sources* (1919) and A. J. Herbertson's introduction led to her employment as a Civil Service examiner

Figure 2.8 The first candidates for the Diploma in Geography, including the only woman, Joan Reynolds, outside the Old Ashmolean (the School of Geography's first home) in June 1901. Credit: University of Oxford Archives.

(she also examined regularly for universities). A prolific author and editor of school geography books, she contributed several articles to *The Geographical Teacher*, including one in its first issue in which she stressed the value of fieldwork using examples provided by her sister who was the headmistress of a girls' school in the Lake District (Walford 2001, 81). She was particularly influenced by Herbertson's ideas about major natural regions (Jay 1965a, 355) as described in her article, 'The regional method of teaching geography' (*The Geographical Teacher* 2 1902–1904, 224). The sixth edition of the Herbertsons' *Man and his Work* (1931) was 'revised and entirely reset with questions and suggestions by J. B. Reynolds' (t.p. verso). Reynolds died in 1950 in France (http://www.pennyghael.org.uk/Reynolds2.pdf) but her much reproduced image continues to epitomize the place that women carved for themselves in the Oxford School of Geography.

Dorothy Forsaith (1885–1972)

Dorothy Forsaith in many ways epitomized early female geography diploma students, being unmarried and a school mistress of lower-middle-class origins who taught at the unfashionable Goldsmiths' College (established for the 'industrial and working classes'), but produced successful and useful geographical publications.

Dorothy Forsaith was born in Wandsworth, London, on 17 March 1885 to Thomas Edward, mercantile clerk, and Helen Lucy, née Shepard, and was baptized into the East Hill United Reformed Church there on 12 April 1885 (London, Non-conformist Registers, 1694–1931, consulted online via Ancestry.co.uk.). She attended Wandsworth High School between 1896 and 1904 (Goldsmiths' College Staff Registers). She was a schoolmistress by 1906 and enrolled in the Society of Oxford Home Students from where she completed the Oxford Diploma in Geography in 1912. She then held a succession of teaching posts, mainly in the north and east of England, before being appointed lecturer in geography in 1919 in the Teacher Training Section of Goldsmiths' College, University of London, founded in 1891 for the education of 'men and women of the industrial, working and artisan classes' (Goldsmiths' College Staff Registers). Forsaith remained there for the rest of her career and participated in the drive to enhance the standing of geography in the school curriculum, as a member of the Geographical Association (she was a member of its council from 1933 to 1936) and as an author of school texts (1920; 1932; 1935; 1951). Although her 1935 book *Other Children's Homes* was aimed at primary school children aged about seven years, most of her work concentrated on children aged 11 to 12 whose interest she engaged with line drawings and photographs, as the reviewers of the first and third editions of *Many Peoples in Many Lands* noted (*Geography* 38(3) (1953), 204; *Geography* 44(1) (1959), 52). Her major publication, *Handbook for Geography Teachers* (1932), was described as an 'indispensable work' and a 'vade mecum' for practitioners (*Geography* 46(1) (1961), 86; *Geography* 51(1) (1966), 82–83), and went through six editions to the mid-1970s. This work originated in the Teaching of Geography Exhibition, held from 3 to 7 May 1932 at Goldsmiths' College in London and opened by Halford Mackinder, of which Forsaith was the chief organizer (Orford 1932). As well as lists of standard textbooks, her book contained references to a wide range of literature, from the poetry of

Shelley ('To a Skylark' and 'Ozymandias'), to travel writing (Gerald Durrell's *Overloaded Ark* and Laurens van der Post's *Venture to the Interior*) and novels (Conrad's *Heart of Darkness*). Although these sections were omitted from the later editions they continued the tradition seen in Fanny Herbertson's *Descriptive Geographies from Original Sources* (Marsden 1988).

Forsaith never married and lived with her sister, Helena Gladys, also a schoolmistress, in Deanhill Road, East Sheen, Richmond, where she died on 23 December 1972, aged 87 (England and Wales, Civil Registration Death Index, 1916–2007 consulted online via Ancestry.co.uk). Like many Oxford diploma students, such as Charlotte Alner Simpson and Gladys Maud Marten (Maddrell 2009, 133–4; 135–7), Forsaith had a long career in geographical education and exerted her influence on the school classroom, as a trainer of teachers, and as the author of teaching texts of considerable longevity. She exemplifies the Oxford school's early impact on the discipline and the space within it for women.

Marie Bentivoglio (1898–1997)

For Marie Bentivoglio (see Figure 2.9), the Oxford geography diploma represented an intellectual and personal turning point in a distinguished

Figure 2.9 Marie Bentivoglio. Credit: University of Sydney Archives (G33_224_1413).

academic career. Her life in general is a reminder that a forward-looking attitude to women's advancement is not always coupled with progressive politics more generally. In 1930 her conservatism was noted in the press ('It is refreshing to meet such a feminine and charming girl whose excellent brain and fund of knowledge have not yet made her into a feminist' (*Sydney Mail* 8 Jan 1930), and it is likely that she sympathized with the fascist regime in power in her native Italy.

Marie Bentivoglio was born in Turin in Italy on 1 August 1898 to Fernando Carlo (Fred Charles) and Esterina (Esther) Cecilia Bentivoglio. The family emigrated thence to Australia when she was just five weeks old (Bentivoglio 2014, which is based on Marie Bentivoglio's notes and from where all biographical details come unless otherwise accredited). Bentivoglio was educated at Fort Street Girls' High School, Sydney, and read for the BSc by specializing in geology, mineralogy and botany at the University of Sydney, graduating in 1919. She was then awarded a Government Science Research Scholarship in the Department of Mineralogy, working on the crystallography of tin ore (her 1921 article is based on this work). She also completed a Diploma in Education from Sydney Teachers' College in 1920. In 1921 she was awarded an 1851 Exhibition Scholarship to enable her to study at the University of Oxford. These scholarships enabled graduates from the British empire to study pure or applied science abroad if they had 'given evidence of capacity for original research'. Bentivoglio was judged to have given appropriate evidence and was only the fourth overseas woman to be elected to one of these scholarships and the first from Australia (Phillips 2001). She matriculated from Somerville College in 1922 to read for the degree of DPhil, Oxford's new 'lower' doctorate established just five years earlier in 1917 (*Oxford University Gazette* 47(1532) 1917, 448–450). Under the supervision of Thomas Vipond Barker and Professor Bowman in the Department of Mineralogy she submitted her doctoral thesis, entitled 'Investigation of the rate of growth of crystals in different directions in relation to their intimate structure' on 2 March 1925. She graduated DPhil that summer, being only the second woman to graduate DPhil from Oxford and the first in the sciences (*Oxford University Calendar* 1920–1926). She published findings from her thesis in 1927 in the highly regarded *Proceedings of the Royal Society of London*.

Bentivoglio's undergraduate studies and the terms of her scholarship ordained that her doctoral research and early publications were in natural sciences, but in Oxford she became a geographer and remained one for the rest of her academic career. While working for her DPhil, she read for the diploma in geography and in January 1924 was awarded the diploma with distinction for the study of her selected region, the Isle of Wight (University of Oxford Register for the Diploma and Certificates in Geography MSS; *Sydney Mail*, 8 January 1930, 34). She embraced the 'new' geography, that is, one with the methods and aims of the sciences (Coones 1989), and vigorously promoted this approach when she returned to Sydney as lecturer in geography in the Teachers' College there from 1926 to 1936. She also lectured in geography at the University of Sydney and was Acting Associate Professor in Geography there between July 1928 and November 1929 in an interregnum between the departure of (Thomas) Griffith Taylor (see *Geographers* 3) and his successor (http://sydney.edu.au/science/geosciences/about_us/au_history.shtml/; accessed 23 July 2018; *Sydney Mail*, 8 January 1930, 34). From 1935 to 1936 Bentivoglio undertook an eighteen-month lecture tour in the United States, speaking on Australian and Italian geography and twice addressing the National Geographic Society of Washington (Biennial Report of the President of the University of Minnesota to the Board of Regents 1934–1936, 49; 274).

Bentivoglio's father, a socialist, was interned in Australia during the Second World War, and she spent much of the war in the United States of America to avoid a similar fate, though her politics were rather different: at Santa Barbara, California, in 1939, she spoke in defence of fascism (*The Crescent of Gamma Phi Beta* May 1939, 25). In 1947 she married Italian Federigo Roberto Eugenio Baldeschi, and in 1949 she became a US citizen. Widowed on 8 November 1954 (Reports of deaths of American citizens abroad, 1835–1974), she returned to Italy in 1964 and died in San Remo on the Italian Riviera on 1 July 1997 (US Social Security Death Index 1935–2014), having been awarded an honorary DSc by the University of Sydney in 1994.

Bentivoglio took the Oxford idea of having 'a special regard to the conditions of life of the inhabitants as affected by and affecting their geographic environment' (Examination Statutes 1916, 58–62) and applied it in popular and well-received lectures on the similarities between the 'young' nations of America and Australia with their desert lands and indigenous peoples.

She embraced the new doctoral path at Oxford, albeit not in geography, and, later, the 'new' geography, pursuing a geographical career on three continents before, during and after marriage. Her geographical publications were largely confined to secondary school teaching (1928; 1931), while her research publications were in technical subjects (1921; 1922; 1927; 1945; 1948). She endowed the Bentivoglio Travelling Scholarship at Somerville College, Oxford, though ironically that college no longer takes geography undergraduates, so the scholarship is closed to students from the subject that marked a turning point in the eponymous donor's life.

Dorothy Mary Doveton (1913–1981)

Dorothy Doveton's life (see Figure 2.10) is one of early distinction in pursuit of geography, which then petered out to be followed by marriage, emigration and involvement in the associational life of her new homeland. Her trajectory

Figure 2.10 Graduates of the Final Honour School in Geography, 1934, including Dorothy Doveton (front row, second from left) with their tutors (front row) Prof. Kenneth Mason, Prof. J. L. Myres and Mr J. N. L. Baker. Credit: University of Oxford Archives.

typifies the path of many women of her generation for whom a geography degree from Oxford was a stepping stone to voluntary work and household responsibilities whereas for their male contemporaries it led to paid employment and in many cases a successful career.

Dorothy Mary Doveton was born on 4 July 1913 in South Africa to Douglas Hodson Doveton, a mining engineer, and his wife Agnes Dorothy (Church of the Province of South Africa, Parish Registers, 1801–2004 consulted online via FamilySearch.org). After he died in Durban in 1920 (South Africa, Natal Province Deaths 1863–1955 consulted online via FamilySearch.org), her mother returned to England where she remarried David Dudley Galbraith in 1923 (England and Wales, Civil Registration Marriage Index, 1916–2005 consulted online via FreeBMD) and the family settled in Bushey, Hertfordshire (UK Incoming Passenger Lists, 1878–1960 consulted online via Ancestry.co.uk). Doveton attended Godolphin School, Salisbury, Wiltshire (Matriculation form, School of Geography archives) and matriculated in 1931 from St Hugh's College. She began reading for the FHS in natural sciences but, after passing the natural science preliminary examination in 1932 (*Oxford University Gazette* 1931–1932, 774), moved into the newly founded FHS of Geography. She was the first person to gain a first class degree (1934) in geography. She won her college's Hurry Prize in 1934 and was president of its JCR (*St Hugh's Chronicle* 1934, 1936). Doveton saw herself as a professional geographer in the making, pursuing research and being a member of the IBG (along with Nora MacMunn) by 1936, just three years after the professional organization was formed (*TIBG* 1936, ix). From 1935 to 1936 she was in Swaziland, as the first of the School of Geography's three Drapers' Company Scholars, where she undertook research for her BLitt into the human geography of what was then a British Protectorate. She published a short article in the *Geographical Journal* (1936, which also circulated as a separately bound pamphlet) and a long piece (110pp.) as a *TIBG* special publication (1937). Powell cites Doveton's work as one of the few geographical publications of the period to have some political content (2015b, 597), and the preface to her 1937 report shows her fully alive to the political context in which she wrote:

> The question of the Protectorates is once again before the public and is likely to rouse controversies within the next year. If this study contributes in any way to the formation of opinion it will serve its purpose and in any

> case … [it] will … partially fill a gap in the geographical literature. … My own studies … were concerned mainly with the geography of the country because I believe that sound policy can only be based on the realisation of the natural possibilities of Swaziland.

This vision of geography as a servant of policy was widespread in Oxford at the time (not just, or even mainly, in the School of Geography) (Symonds 1986; Louis 2013). Certainly it was one shared by Kenneth Mason, Oxford's newly appointed professor of geography, Doveton's thesis supervisor, and advocate of 'applied geography' (Mason 1938; Hansen 2004). Doveton's version of it was primarily humane: she thought that policy should be determined by the interests not of the rulers (here Britain or South Africa) but the ruled, that is, the Native and European inhabitants of Swaziland (in that order) (1937, 1), something that was problematic since she thought geography made union with South Africa rational, but cut across the wishes of 'the Natives' who, more perspicacious than she as it turned out, mistrusted South Africa's 'Native Policy' (1937, 3). Her treatment of her subject was inevitably of its time, with its discussion of Britain's 'moral obligation' to a 'subject race' (1937, 3), but is thorough and enlightened within its own frame of reference and does not hesitate to criticize imperial policy (1937, 41). In her shorter 1936 paper she is even more clearly writing in a politicized way and does not shy away from taking issue with the conclusions of official reports (1936, 331). She was unlucky enough to have her work reviewed by Samuel Baker (in *Africa* (1939, 12(3), 386–7) and *Geography* (1939 24(1), 62–3) who was trying to make a name for himself as the authority on the geography of East Africa (Ominde 1971) and evidently did not like the competition that Doveton represented. His reviews were carping though not dismissive.

Doveton probably chose Swaziland because of her family connections and because Swaziland's political future demanded 'immediate attention' (1937, 3), and it was small enough to make its study practicable for a BLitt thesis. Doveton appears not to have published anything else, perhaps stung by Baker's reviews, but her interest in Africa was genuine. On 11 December 1937 at Holy Trinity Church, Brompton, London, she married fellow South African Patrick Cecil Dicey, a surveyor. By 1941 they had moved to Bulawayo, in Southern Rhodesia, where their three children were born (Institute of Heraldic and Genealogical Studies, Andrews Collection, Newspaper Index Cards, 1790-1976; consulted online via Ancestry.co.uk). She was appointed Member of the

Order of the British Empire in the 1961 birthday honours list for 'social welfare and cultural services in the South Western part of Southern Rhodesia' (*London Gazette* 2 June 1961 (42370), 4165). Following the death of her first husband in 1966 she married John Fleming on 5 August 1978 (*St Hugh's Chronicle* 1978–1979). She reported to the *St Hugh's Chronicle* for 1980–1981 that she was a board member of the National Free Library, Bulawayo, Zimbabwe, a representative of the National Federation of Women's Institutes of Zimbabwe, and an Executive Member of the National History Society. She died of cancer on 18 January 1992 in Plettenberg Bay, Western Cape Province, South Africa. The society for geographers at St Hugh's College was named after her, but it lapsed when in 2005 the last geography undergraduates at the college sat for their final examinations (*St Hugh's Chronicle* 2004–2005, 8).

(Isobel) Moira Dunbar (1918–1999)

Moira Dunbar was a graduate of the Oxford school who achieved very considerable scientific distinction. She is included here as an exemplar of the international reach of women at the school (like Doveton she moved within the British imperial sphere) and of how marriage, or in her case, lack of marriage very substantially circumscribed women's careers (unlike Doveton she did not marry and was thus free to pursue a scientific career). She deserves a full memoir in *Geographers*: the brevity of her biography here reflects simply the limited role the Oxford School of Geography played in her life.

(Isobel) Moira Dunbar was born in Edinburgh on 3 February 1918, the daughter of William John Dunbar, who was a well-known sheriff and advocate at the Scottish bar. She attended Cranley School for Girls in Edinburgh (Anon 1999) before reading for a BA in geography at St Anne's College and the School of Geography, graduating in 1939 (*Oxford University Gazette*, 69, 1939). She emigrated to Canada in 1947 and became a particularly distinguished Arctic scientist, specializing in sea ice research. With its obvious defence applications and during the Cold War, this was a particularly challenging field for a woman and her early efforts to sail on Canadian Royal Naval vessels to the Arctic were rebuffed ('they regarded me as some sort of cross between a delicate flower and a dangerous disease' (Thomas 2000). Dunbar persevered and her research was recognized by her receipt of the Meteorological Service of Canada's Centennial Award (1971); the Royal Canadian Geographical Society's Massey

Medal (1972)[4] and the Royal Canadian Navy Admirals' Medal (1988), the sole female recipient of this medal to date. She was appointed Officer of the Order of Canada (1977) and was elected Fellow of the Royal Society of Canada. She was head of the Royal Canadian Geographical Society (Thomas 2000; Royal Canadian Navy Admirals' Medal recipients 2018; Hulbe, Wang and Ommanney 2010; Vincent and Simpson-Lewis 2009; Anon 1999; Rowley 2008; http://www.science.ca/scientists/scientistprofile.php?pID=346 2015).

Women at Oxford's School of Geography: a brief statistical account

The diploma and certificate in geography

From their inception in 1901 to their demise in 1939, Oxford's diploma in geography and to a lesser extent the shorter certificate proved popular with women: 168 women and 100 men took the diploma, while 39 women and 108 men took the certificate, and 18 men took the certificate in surveying. However, between 1920 and 1933 only one woman received the Herbertson Prize for 'exceptional merit in the "Geographical Study of a Selected District"', against seven men; thereafter the prize was associated with the FHS course. Of the short course students, 38 women but only 7 men arrived in Oxford with an external degree, most of the other men being BA Oxon. Following the introduction of the FHS in geography, the diploma and certificate declined in popularity with women in particular: between 1932 and 1938 only 12 women (7 for the certificate and 5 for the diploma) enrolled for these courses whereas 52 men (41 for the certificate and 11 for the diploma) did so. As discussed above, diplomas were particularly suited to school teachers, and this accounts for their popularity with women graduates from outside. Table 2.1 shows several early candidates from Cherwell Hall, a women's teaching training establishment that was never part of the University (though its building was later incorporated into a women's college: St Hilda's), something which again shows women and geographers on the edge of University life. Of the more mainstream women's societies at Oxford, the one that took most diploma students was the Society for Home Students, a society of students unable to afford the fees of other halls.

Table 2.1 Women diploma and certificate students at the Oxford School of Geography 1900–1940

Date	Name	College or Society
1901	Joan Berenice Reynolds	Geography School
1904	Nora Eileen MacMunn	
1905	Geraldine Wooldridge	
1906	Mary Priscilla Weston	
1907	Margaret Hannah	Lady Margaret Hall
1908	Eva Germaine Rimington Taylor	
1908	Ellen Jenifer Rickard	Somerville
1908	Eva Dodge	Cherwell Hall
1908	May Hockley	OHS
1908	Alice Emma Lee	
1908 / 22	Lydia Margaret Livermore	OHS
1908	Edith Jane Wilford	St Hugh's Hall
1909	Madelaine Catherine Fripp	OHS
1909	Margaret Muriel Grey	Cherwell Hall
1909	Marie Hermione Ernestine Wulff	
1909	Laura Black	OHS
1909	Ella Louise Gentry	OHS
1909	Evelyn Mary Oswald	OHS
1909	Gladys Pugh	St Hilda's Hall
1909	Margaret S.F. Cope	Cherwell Hall
1910	Jessie McWean	Cherwell Hall
1910	Charlotte Alner Simpson	OHS
1910	Lavinia Mary Hardy	OHS
1910	Ethel Frances Henson	OHS
1910	Harriet Annie Hartley	Somerville
1910	Ginevra Theodosia Henslow	OHS
1910	Edith Ainsbury Johnson	OHS + Cherwell Hall
1910	Phoebe Julia Oxley	OHS
1910–11	Edith Marion Coulthard	OHS
1910	Lille Stephanie Loos	OHS
1910	Ellen MacDonald	OHS
1911	Jessie Latto Lowson	OHS
1911	Catherine Moffatt	OHS
1911	Ethel Reece Wilkinson	OHS
1911	Sara Ackroyd	OHS
1911	Alice Noele Porter Gregg	OHS
1911	Hester [Carter] Lamb	OHS
1911–12	Evelyn May Graham Sandberg	OHS
1911–12	Annie Templeton Anderson	OHS
1911–12	Mary Edith Coates	OHS
1911	Effie Dawson	OHS
1911–12	Dorothy Mary Forsaith	OHS
1911–12	Elfrida Margaret Robin	OHS

(*Continued*)

Table 2.1 (*Continued*)

Date	Name	College or Society
1911–12	Phoebe Mary Swift	OHS
1912	Elizabeth Pitcairn Hogg	Cherwell Hall
1912	Gladys Maud Marten	OHS
1912	Lilian Elizabeth Auld	St Hugh's
1912	Hilda Howell	OHS
1912	Hilda Annie Stephanie Perry	OHS
1912–13	Vera Katherine Isaac	OHS
1912	Marjorie Laidler	Cherwell Hall
1913	Mabel Mary Barker	St Hilda's
1913	Agatha Booker	OHS
1913	Margaret Elizabeth Gray	OHS
1913	Annie Halstead	Cherwell Hall
1914	Blanche Hosgood	OHS
1914	Dorothy Donovan Adam	Somerville
1914	Grace Churchill	OHS
1914	Margaret Arthur Harvey	OHS
1914	Mary Winifred Newton	Somerville
1914	Evelyn Decima Willes	OHS
1914–15	Frances Geraldine Hailes Coster	OHS
1914–15	Irene Jane Gregory	OHS
1914	Muriel Clare Hasluck	OHS
1914–15	Violet Helen Truman	St Hugh's
1915	Ethel N. Allen	OHS
1915	Mary Josephine Connaughton	OHS
1915	Adeline Winifred Vaughan	OHS
1915	Beatrice Emma Miles	OHS
1915 / 22	Dorothy Richmond Bagley	OHS / St Hugh's
1915	Margaret Hailley Chambers	Somerville
1915–16	Edith Cross	OHS
1915	Eileen Isobel Glenday	St Hugh's
1915–16	Mary Kingdon Heslop	OHS
1915 / 23	Vivienne Cecilia Horne	St Hugh's
1915–16	Katherine Mabel Kempthorne	St Hilda's Hall
1916	Catherine Elizabeth Clegg	OHS
1916–17	Ethel Jane M. (Elsie Jill) Haynes	OHS
1916	Joan Dorothea Hughes	OHS
1916	Mary Levin	St Hugh's
1916	Lilian Mary Meloille Wanstall	OHS
1916	Gertrude Lilian Winser	Somerville
1916	Margaret Dora David	OHS
1916–17	Kathleen de la Mare	Somerville
1916–17	Lorna Margaret Bertha Howell	St Hilda's
1916–17	Edith Frances Peto	OHS
1917	Gladys Baines	OHS
1917	Elsie Mary Butterworth	St Hugh's

Table 2.1 (*Continued*)

Date	Name	College or Society
1917 / 21	Margaret Powell	OHS
1917–18	Joyce Mary Baker	Somerville
1917–19	Muriel Cranfield	OHS
1917–18	Dorothy Marion Cross	OHS
1917–18	Dorothy Hillkirk	OHS
1917–18	Gladys Boyd Hurry	St Hugh's
1918	Margaret Eveline Doyle	OHS
1918	Elizabeth Ethel Feather	OHS
1918–20	Philippa Winifred Drew	OHS
1918 / 23	Florence Marjorie Exton	OHS
1918–19	Dorothy Fosbrooke	OHS
1918–20	Violet Ethel Latham	St Hilda's
1918	Annie Gertrude Rayfield	OHS
1919	Kathleen Mary Locker	St Hilda's
1919–20	Charlotte Elizabeth Lilias MacLeod	OHS
1919	Frances McKechnie	OHS
1919	Mary Wilton Parkin	OHS
1919–20	Lorna Mary Emma Vincent	OHS
1919	Hilda Mary Watson	OHS
1920	Katharine Maud Atkinson	OHS
1920–21	Mary McInnes	OHS
1920–21	Gertrude Louisa Newby	OHS
1920	Jeanie Christina Stratton	OHS
1920 / 22	Mabel Daisy Biddick	OHS
1920	Edna Brown	OHS
1920–21/2	Margaret Pudsey Dawson	OHS
1920–22/3	Kate Doreen Freeman	OHS
1920–21/2	Freda Jarvis	OHS
1920–21/2	Phyllis Johns	OHS
1920–22	Kathleen Mary Pond	OHS
1920	Hilda Bessie Pritchett	OHS
1920–21/2	Dorothy Rose Robson	OHS
1920–21/2	Elizabeth Marjorie Sheavyn	OHS
1920–21/2	Marion Kathleen Trethewey	OHS
1920–21/2	Phyllis Wray-Bliss	OHS
1921–22/3	Agnes Clara Booth	OHS
1921–22/3	Isobel Gertrude Goddard	OHS
1921–21/2	Irene Helen Lowe	OHS
1921–21/2	Ethel Agnes Mackay	OHS
1921–21/2	Eva Mabel Mitchell	OHS
1921–22	Lydia Emma Shaw	OHS
1921–21/2	Florence Sarah Slatter	OHS
1921	Edith Murial Blackburn	OHS
1921–24 /28	Bertha Mary Broadhurst	OHS
1921–21/2	Mary Rebekah Lambert	OHS

(*Continued*)

Table 2.1 (*Continued*)

Date	Name	College or Society
1921–22/3	Clara Margery Lawrence	St Hilda's Hall
1921–21/2	Evelyn Hattie (Hettie) Lovett	OHS
1922	Dorothy Agnes Edwards	OHS
1922–24	Mary Cecilia Howell	OHS
1922–24	Ethel Kyte Larwill	OHS
1922–23	Lilias (or Lilian) Wright	OHS
1922	Barbara Maud Fleur Dundas	OHS
1923–24 / 25	Marie Bentivoglio	Somerville
1923–24	Helena Mabel Edwards	Somerville
1923–24	Alice Maud Lunn	OHS
1923–24	Cecilia Katherine Gillies Smith	OHS
1923–24	Hilda Margaret Tilley	OHS
1923	Lily Graham	OHS
1923–25	Margaret Irene Alice Hastings	OHS
1923–25	Carol Evelyn May Rygate	St Hugh's
1923–25	Slyvia Storrs Fox	OHS
1923	Winifred Lilian Walters	OHS
1924–6	Evelyn Florence Beck	OHS
1924–5	Irene Jenny Josephine Carlebach	OHS
1924–6	Muriel Hermione Cooper	OHS
1924–5	Ellen Annie Moore	OHS
1924–6	Marjorie Dieudonnée Wood	OHS
1924–6	May Evelyn Webb	OHS
1925–27	Kathleen Kennedy Cooke	OHS
1925–6	Mona Kathleen Jones	OHS
1925–6	Gladys Nash	OHS
1925–6	Hawys Nesta Rhys Jones	OHS
1925–6	Mary Ellen Baron Russell	St Hugh's
1925–6	Margaret Mary Shepherd	OHS
1925–27	Ida Evelyn Fletcher	OHS
1925	Beatrice Marie Sellers	OHS
1925–27	Joan Summers	OHS
1926–28	Kathleen Emily Babbs	St Hugh's
1926–28	Kathleen Marianna Hogg	OHS
1926–27	Margaret Edith Kinhead	OHS
1926–27	Phyllis Gwendoline Moss	St Hugh's
1926–28	Sybil Gertrude Notley	OHS
1926–28	Eleanor Pocock	OHS
1926–28	Elizabeth Ellen Shaw	OHS
1926–28	Elaine Wade	OHS
1926–27	Flora Christian Welch	St Hugh's
1926–28	Rosamund Carus	OHS
1926–28	Ruth Cecilia Shaw	OHS
1927	Agnes Isobel Symmington	OHS
1927–29	Kathleen Mary Cliff	OHS
1927–29	Mabel Elizabeth Cole	OHS

Table 2.1 (*Continued*)

Date	Name	College or Society
1927–29	Nellie Louise Crowe	OHS
1927–28	Ethel Mildred Hastings	OHS
1927–29	Nancy Lomax Pearce	OHS
1928–30	Mary Banbury	OHS
1928–30	Wilhelmina Mary Hogan	Somerville
1928/30	Edith (or Effie) Lyall	
1928–30	Pamela Edith Matthew	OHS
1929–31	Olive Henderson	OHS
1929–31	Barbaras Swanson Hunt	OHS
1929	Mary McCrinick	OHS
1929–30	Mary Tilbrook	St Hilda's
1929	Elsie May Wright	OHS
1930–31/33	Elizabes Agnes Anderson	OHS
1930–31	Ada Dorothea Catharine Fenning	OHS
1931–32	Mary Hepple Dickinson	OHS
1931–32	Edith Maud Way	
1932–33	Joyce Theodora Margaret Biggar	
1932–33	Edith Finlay Currie	OHS
1932–33	Janet Mary White Lennox Scott	OHS
1932	Winifred Mary Thorpe	OHS
1933	Dorothy Blanche Morgan	St Hugh's
1934	Edith Portia Hobbs	Somerville
1935–36	Fedelma Nora Donnelly	
1936	Monica Vibert Moon	Somerville
1936	Nora Gilbertson	Somerville
1938	Joan Hulda Johnson	St Hilda's
1938	Enid Adela Malin	St Hilda's
1938	Joan Elizabeth Rachel Salter	St Hugh's

OHS: Oxford Society for Home Students.

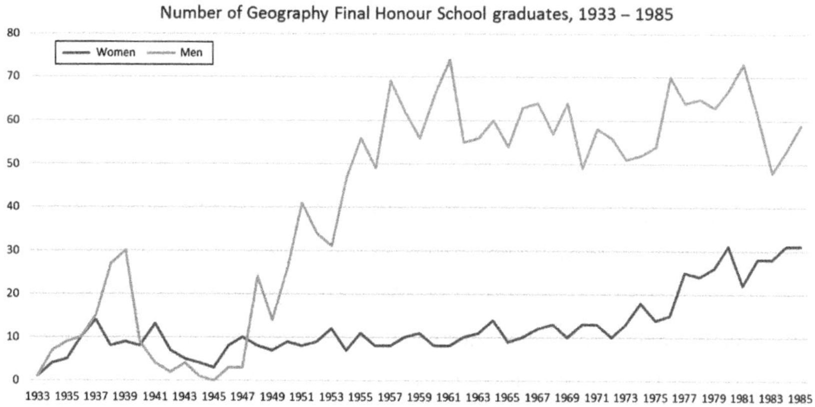

Figure 2.11 Women and men FHS graduates 1933–1985 at the Oxford School of Geography.

The Final Honour School in geography

Introduced in 1932, the FHS had its first two students graduate in 1933 (see Figure 2.11 and Table 2.2). Of the two, the woman, **Janet Mary White Lennox Scott**, in many ways epitomized early women geographers: from the Society of Home Students, a graduate from another university, and holder of Oxford's certificate and diploma. Other than in the exceptional period of the Second World War, women undergraduates were outnumbered by men, and increasingly so. From the early 1970s the number of women geography undergraduates rose steadily, such that by the 1980s around 30 women per year graduated BA, compared with an average of 60 men per year. A total of 664 women took the geography FHS course in the 50 or so years after its introduction, compared with 2,190 men (*Oxford University Gazette*, 1932–1985).

Graduate students

Before 1920 women geography graduate students were examined by geography staff but awarded either a Certificate of Letters in modern history or a Certificate of Science in natural science. After women could matriculate (1920 onwards) the graduate degrees of BSc and BLitt were opened to women. All the earliest female research students took the diploma in geography before reading for the BLitt or BSc, the first being **Frances Geraldine Hailes Coster** who took the diploma in 1914/15, and was admitted as an advanced student in 1916 and awarded the Certificate of Letters in March 1919 (*Oxford University Gazette* 49 1918–1919, 358, 659), before matriculating and graduating BLitt in October 1920 (*Oxford University Gazette* 51 1920–1921, 92, 134). Coster was one of nine women who completed the diploma and/or certificate before reading for the BLitt or BSc in geography; after 1933 most graduated BA before reading for BLitt or BSc, or, exceptionally, the DPhil.

The first geography DPhil student was **En-Lan Liu** (*fl.* 1923–1950), who graduated BA from Ginling College, a women's college of Christian foundation in Nanking/Nanjing (1925), and MA from Clark University, Worcester, United States of America (1925). She already had at least one article in print when she matriculated from St Hilda's College in November 1938. Her thesis was on rainfall in China and she graduated DPhil in absentia in November 1940 (*Oxford University Gazette*, 69 1938–1939, 139;

Table 2.2 Women FHS students at the Oxford School of Geography 1933–1950

Date	Name	College or Society	Class
1933	Janet Mary White Lennox Scott	OHS	3
1934	Dorothy M. Doveton	St Hugh's	1
	A. Craig	OHS	2
	Winifred Mary Thorpe	OHS	2
	Winifred Mary MacDonald Troup	St Hugh's	2
1935	Margery A. Clerk	St Hugh's	2
	Constance M. Exley	St Hugh's	2
	Margaret R. Wilson	St Hugh's	2
	Ena W.M. Coleman	OHS	3
	Mary P. Lee	St Hugh's	3
1936	Joan R. Ardran	LMH	2
	Pauline M. Brentnall	St Hugh's	2
	Nancy M.L. Field	St Hugh's	2
	Anastasia Swinburne	OHS	2
	Kathleen I. Teasdale	St Hugh's	2
	Frances A.A. Deas	St Hugh's	3
	Katherine G. Harvie	OHS	3
	Olive J. Lavender	OHS	3
	Margaret E.E. McDougle	St Hugh's	3
	Cynthia J.L. Stenhouse	OHS	3
1937	Louisa M. Bankes	OHS	2
	Frances E. Barendt	LMH	2
	Margaret J. Bulman	OHS	2
	Ruth Langhorne	OHS	2
	Frances W.L. Moore	LMH	2
	Eileen M. Page	OHS	2
	Nora C. Shaw	St Hugh's	2
	Helen J. Southern	St Hugh's	2
	Margaret P. Walker	OHS	2
	Patricia P. Backhouse	LMH	3
	Kathleen M. Cane	St Hugh's	3
	Adelaide J. Ferguson	LMH	3
	Mary E. Odell	OHS	3
	Dorothy Payne	OHS	3
1938	Joan E.M. Campbell	OHS	1
	Elizabeth A. Aste	Somerville	2
	Diana Coxhead	OHS	2
	Patricia M. Dolby	OHS	2
	Mary A. Lovegrove	OHS	2
	Edith M. Ness	OHS	2
	Lucy J. Becher	LMH	3
	Isabella J. Quixley	OHS	3
1939	Mary Marshall	LMH	1
	Lilian M. Benger	Somerville	2
	Marguerite D. Emmens	OHS	2

(*Continued*)

Table 2.2 (*Continued*)

Date	Name	College or Society	Class
	Hannah M. Gilmour	St Hugh's	2
	Constance I. Kahn	St Hugh's	2
	Joan Robinson	OHS	2
	Jean L. Crichton-Miller	OHS	3
	Isobel M. Dunbar	OHS	3
	Margaret A. Gregory	OHS	3
1940	Diana M.R. Crowfoot	Somerville	2
	Faith Y. Ede	OHS	2
	Dorothy E. Penny	St Hugh's	2
	Audrey M. Downie	St Hugh's	3
	Barbara E. Griffiths	OHS	3
	Pamela G. Schiele	Somerville	3
	Joan M. Willson	OHS	3
	Joan Wiltshire	St Hugh's	3
1941	Bernice D. Baker	LMH	1
	Sheila F. De Sa	St Hugh's	1
	Mary I. Anderson	LMH	2
	Prudence B. Davies	St Hugh's	2
	Dorothy Fowkes	OHS	2
	Joan O. Harries	St Hugh's	2
	Myrtle B.D. Morrison	LMH	2
	Elizabeth G. Wilson	LMH	2
	Joan M. Godfrey Isaacs	OHS	3
	Evelyn M. Pelly	OHS	3
	Helen F. Pickard-Cambridge	OHS	3
	Isabel Watson	OHS	3
	Olivia E. Wright	OHS	4
1942	Eileen M. Stephenson	LMH	2
	Eleanor C. Vollans	St Hugh's	2
	Stella M. Mortimore	OHS	3
	Barbara J. Sanderson	St Hugh's	3
	Ynys D. Scott	St Hugh's	3
	Lettice M. Wigram	LMH	3
	Joan H. Wilson	OHS	4
1943	Kathleen M. MacIver	LMH	2
	Margaret B. Pritchard	St Hugh's	2
	Mary E. Denham	St Anne's	3
	Nora Elliott	St Hugh's	3
	Marjorie E. Sutton	LMH	3
	Marguerite A. Lloyd	LMH	2
	Margaret S. Plowman	St Hugh's	2
	Margaret A. Powell	LMH	3
	Thérèse M. Walter	St Anne's	3
	Jean M. Whiting	St Anne's	3
	Joan H. Woodward	St Anne's	3
	Barbara W. Brown	St Hugh's	2
	Olwyn D. Reynolds	St Hugh's	2

Table 2.2 (*Continued*)

Date	Name	College or Society	Class
	Joan N. Norton	St Anne's	3
	Pamela J.C. Salmon	LMH	3
1944	Mrs Joyce E. Chaplin	St Anne's	2
	Betty Gill	St Anne's	2
	Margaret D. Prior	St Anne's	2
	Mary N. Blake	St Anne's	3
	Margaret M. Taylor	St Anne's	Aegrotat
	Doreen M. Harvey	St Hugh's	3
	Marguerite A. Lloyd	LMH	2
	Margaret S. Plowman	St Hugh's	2
	Margaret A. Powell	LMH	2
1945	Pamela C.M. Creek	LMH	2
	Alice M. Goodrich	LMH	2
	Jocelyn M.E. Fortescue-Foulkes	St Hugh's	4
1946	Mary G. Hardy	LMH	2
	Audrey M. Murphy	LMH	2
	Jean W. Potter	St Anne's	2
	Mary L. Woodward	St Hugh's	2
	June C. Blomfield	St Hugh's	3
	Sylvia A. Peto	St Anne's	3
	Heather R. Webster	LMH	3
	Doreen M. Harvey	St Hugh's	4
	Sonya Gregory	LMH	2
1947	Mrs Jane M. Barbour	St Hugh's	Unclassified honours
	Jeanne Brassington	St Hugh's	2
	Margaret Drury	St Anne's	2
	Yvonne E. Mead	St Hugh's	2
	Muriel M. Rees	St Hugh's	2
	Gwendolyn J.E. Berry	St Anne's	3
	Joan C. Heslop	St Hugh's	3
	Patricia M. Lindsay	St Hugh's	3
	Mary P. Stott	LMH	3
	Dorothy M. Walker	St Anne's	3
	Mrs Gwenda J. Whitty	St Hugh's	3
	Sonya Gregory	LMH	2
1948	Ethel E. Britton	St Anne's	2
	Evelyn A. Britton	St Anne's	2
	Mary K. Kinton	LMH	2
	Margaret M. van der Plank	LMH	2
	Helen M. Wallis	St Hugh's	2
	Jean M. Whiting	St Anne's	2
	Jean M. Bullen	St Hugh's	3
	Yvonne M. Lehmann	St Anne's	3
1949	Patricia M. Brassington	St Anne's	2
	Patricia M. Devlin	LMH	2

(*Continued*)

Table 2.2 (*Continued*)

Date	Name	College or Society	Class
	Anne H.H. Moore	LMH	2
	Marjorie P. Paine	St Hugh's	2
	Gilian K. West	St Hugh's	2
	Moira K. Hennessey	St Anne's	3
	Margaret M. Smith	St Anne's	3
1950	Dorothy Armitage	Somerville	2
	Barbara J. Missen	St Hugh's	2
	Audrey Nugent	St Hugh's	2
	Louisa B. Darley	St Anne's	3
	June M. Eastwood	St Anne's	3
	Mary M. Evans	St Anne's	3
	Nora V. Lodwick	St Anne's	3
	Edna M. Walker	LMH	3
	Doris M. Imrie	St Anne's	4

Source: *Oxford University Gazette*, 1933–1950.
OHS: Oxford Society for Home Students.
LMH: Lady Margaret Hall.

71 1940–1941, 129). Faced with a lack of systematic meteorological data and the impossibility of compiling any, given the political situation, her thesis was largely given over to analysing such few data as existed, drawing comparisons with other parts of the world and making recommendations for the future. She returned to a tenured post at Ginling College, and once there published articles and gave lectures to soldiers to help foreign military personnel in China understand the country – in fact, producing on a much smaller scale variants of the Naval Intelligence Handbooks that the school was producing while she was there – but she added hopes of international peace and understanding, which are moving given what her college suffered in the Nanking massacre of 1937–1938 (e.g. Liu 1937; 1946). She was active in geographical publication until 1952, publishing mainly in US teaching and other geography periodicals. Her familiarity with the US higher education system perhaps explains why she applied so early on for the Oxford DPhil (still a novelty in Britain), and her graduation anticipated by nine years that of the first successful male doctoral student, Robert Percy Beckinsale (1908–1998; see *Geographers* 22), who graduated DPhil in 1949, though, of course, the war made the whole period atypical. From 1970 the DPhil was the most common graduate qualification, although a handful of women graduated MSc or MLitt (see Table 2.3 for Oxford geography graduate students).

Table 2.3 Women graduate students at the Oxford School of Geography 1920–1950

Date	Name	College or Society	Degree	Supervisor	Thesis title
1920	Frances Geraldine Hailes Coster	OHS	BLitt		Vale of Somerset: a study of economic development and types of settlement in relation to physical characteristics
1922	Charlotte Alner Simpson	OHS	BSc		Regional geography of the Cotteswolds
1923	Mary McInnes	OHS	BLitt		The human geography of Sheffield and the surrounding district
1924	Elsie Jane MacDonald Haynes	OHS	BLitt		Human geography of the English pottery industry
1924	Edith Cross	OHS	BLitt		The geography of the milk supply of London down to 1921
1924	Lorna Mary Emma Vincent	OHS	BSc		Landforms of the western slopes of the Pennines
1927	Margaret Powell	OHS	BSc		Geographical survey of the Mendip Hills showing their effect upon human activities in the region
1929	Agatha Booker	OHS	BLitt		A geographical study of wheat and the wheat trade in Europe
1930	Kathleen de la Mare	Somerville	BLitt		The place of the olive in the economy of the Mediterranean region
1933	May Evelyn Webb	OHS	BLitt	N. MacMunn	A comparison between the geographical bases of the rural economic life of the Vale of the White Horse and that of White Horse Hills
1934	Dorothy M. Doveton	St Hugh's	BLitt	K. Mason	The Human Geography of Swaziland
1937	F.E. Barendt	LMH	BSc	Mr Gullick	The economic geography of the linen industry in Ireland

(Continued)

Table 2.3 (*Continued*)

Date	Name	College or Society	Degree	Supervisor	Thesis title
1938	En-Lan Liu	St Hilda's	DPhil	P.M. Roxby	A study of the rainfall fluctuations in China: their causes and effects
1941	S.F. Desa	St Hugh's	BSc	E.W. Gilbert	The regional geography of educational admission in England and Wales
1946	E.C. Vollans	St Hugh's	BLitt	E.W. Gilbert	The county borough of Derby, in relation to the surrounding region
1947	J. Carter	LMH	BLitt	M. Marshall (C.S. Orwin from 1948)	A functional classification of settlement in Breckland
1947	J. Higginson	St Hugh's	BLitt	C.S. Orwin	A geographical survey of fruit growing in the Vale of Evesham and the Wisbech area
1947	M.M. Rees	St Hugh's	BLitt	R.P. Beckinsale	Rural settlement in North Wales since the late eighteenth century
1948	V.A. Janaki	LMH	DPhil	J.N.L. Baker	Population and settlement in Kerala
1948	H.M. Wallis	St Hugh's	BLitt	R.A. Skelton	The influence of physical factors on the exploration of the South Pacific, 1519-1770
1950	M.E. Webb	St Anne's	DPhil	R.W. Steel	Jamaica – an island study

Source: *Oxford University Gazette*, 1920–1955.
Note: Dates from 1920 to 1930 indicate year of graduation; dates from 1933 to 1955 indicate year of matriculation.
OHS: Oxford Society for Home Students.
LMH: Lady Margaret Hall.

Conclusion: Oxford's School of Geography as an accidental site for women's advancement

'In Oxford....but not of Oxford': thus Janet Howarth echoed an 1896 comment on the place of women at Oxford (2000b) and it is worth asking whether the same was true of women in the School of Geography: Were they an essential part of the place, helping to shape its contemporary culture and future direction, or did they have to conform to, and exist in the margins of, structures, curriculums, social structures and teaching methods designed by and for men?

It is clear from the pronouncements of the school's leaders quoted earlier in this essay that their desideratum was to provide specialist, advanced geographical training for those destined for a career in public life. Public life was what the school's first Reader, Halford Mackinder, ultimately chose over geography as a career (he was an MP for fourteen years, more briefly a diplomat, and always concerned with imperial affairs and geopolitics (Kearns 2009)), and it shaped the priorities of the school's first statutory professor, Kenneth Mason, who was an imperial surveyor and explorer before his appointment to Oxford and whose most significant work at Oxford was as leader of the Naval Intelligence Handbooks project. The public life that Mackinder and Mason valued was exclusively male: the army and navy, surveying, national and imperial politics, the civil service. They had no discernible interest, for example, in local government work in which women were active as public servants. The school's first personal professor, A. J. Herbertson, was more interested in scientific enquiry than public service, but, shaped by his early training in meteorology, ideally placed geography among the natural science schools, which were very largely male.

Had these men had their way geography's story at Oxford would have been very different and women would have been wholly marginal to it; but they largely could not, notwithstanding representations of the Oxford school as a servant of empire (Symonds 1986; Symonds 2000) – a characterization that treats men's experience alone as representative. Mackinder's, Mason's and Herbertson's subject remained poorly funded and on the margins of Oxford life: that is, in Oxford but not very convincingly of Oxford. And in these circumstances women's economic power won them a central place in the school. Women, in possession or anticipation of an income as a schoolmistress, could pay the fees for courses that served their need (or their families could pay them in the hope of making their daughters self-supporting), and the fees made their way to the school's staff in a surprisingly direct way. O. G. S. Crawford, junior demonstrator, recalls how after the Oxford geography summer schools, Herbertson would do the accounts (fees minus expenses) and share out the profits to the staff (1955, 62). Crawford described himself as 'hard up' (1955, 48) and the Herbertsons' summer courses for school teachers and relentless production of text books speak of a need to make money, not simply of dedication to their discipline. (Herbertson's will was proved in 1916 at a mere £2,224 14s. 5d. and it was obvious that the fact of being a spouse and parent affected his professional choices as much as they did Fanny's (Baigent 2004a).) Such school books kept the University's press in profit from the 1870s, and they did the same for their many authors (Fraser 2014, 443). Women with their salaries or with the support of enlightened families, then, carved a place among the students at the school and, then armed with their newly minted geographical qualifications took their place on the teaching staff: as we have seen, most of the demonstrators in the 1910 summer school were female and had been students at the school.

A turn towards school teaching in itself was not enough to keep Oxford geography afloat. The school had to train its sights not just on the great boys' public schools that fed into mainstream Oxford disciplines such as classics and divinity: being marginal in Oxford, geography had to reach across the gender divide into girls' schools, down the age range into primary schools, and down the social scale into minor public schools and state schools. Oxford geographers wrote for all of these groups and trained women and men to teach in all of these settings. There are again interesting parallels between histories of geography and of the University press, long rather marginal to the University's perception of itself, if not to its balance sheet. Market demands forced the press's educational publishing to look beyond the market of the best boys' public schools to provision for a whole range of other schools (Fraser 2014, 444–5), just the sort of school that Oxford's geographers were also training their sights on.

The result of these facts was an early Oxford geography school that, for all the male posturing evident in early histories, was remarkably attuned to women's needs and which gave women a stage on which they acted as men's equals. A very nice memoir by a Mr McGregor on the first summer school describes with excitement the first lecture as he met his classmates 'half being ladies' (175) and suggests that women and men shared the experience as equals. Women as well as men ended up looking 'not very presentable' as they rummaged in quarries for fossils, women as well as men identified the fossils, and the repartee of 'an American lady' at the writer's expense enlivened the final day (Carter and McGregor 1901–1902). O. G. S. Crawford a decade later remembers, 'The majority of the students at the School of Geography were women whom, as Junior Demonstrator, it was my duty to teach.....My job was to teach geography to a small group of students, most of them women' (1955, 70–74). The statistics on the Oxford geography short courses support their perceptions: at least in the days before the FHS was founded, women students were certainly in geography and of geography.

Women's place in the school was both the consequence and the cause, however, of the fact that geography was in Oxford but not convincingly of Oxford. The school needed them to keep numbers up, but the more women were involved, the less geography resembled the disciplines with which its leaders hoped to carve it an equal place. Geography's marginality at Oxford is generally told in disciplinary or structural terms: geography lacked the social cachet of Oxford's traditional schools such as classics, and had to wait more than 30 years before it featured as an FHS and longer still before its tutors all had college fellowships. More obvious, if largely unremarked, was how socially marginal many geographers were. Mr McGregor's summer school memoir makes this clear: the summer school was priced in guineas, the currency of the upper-middle class, but he was counting the shillings and probably also the pennies when he attended. He also wrote wistfully of the 'je ne sais quoi' of the Oxford atmosphere – something that he regarded himself as glimpsing from the outside (Carter and McGregor 1902, 175; *The Geographical Teacher* 1,3 1901–1902, 124). 'A large number' of summer school students benefitted from local authority bursaries (Gilbert 1972, 193), that is, they taught at schools for the bulk of the population rather than the elite. Similarly, Steel suggests that the School of Geography had links with grammar schools (for the middle- and lower-middle classes) rather than with (elite) public schools, which had an increasing grip on the rest of Oxford (1987, 62; Honey and Curthoys 2000). As with the summer schools, so with the term time courses: upper-middle class O. G. S. Crawford told an anecdote about geography's marginal social position from the reverse class perspective to Mr McGregor's. He recalled in his memoirs that when he told his college tutor he was giving up 'Greats [classics] for geography 'it was like a son telling his father he had decided to marry a barmaid. There was nothing he could do about it and he knew it. ... Greats is for brainy people. Going from Greats to Geography was like leaving the parlour for the basement, one lost caste but one did see life' (Crawford 1955, 61, 78).

If early Oxford women students were in and of Oxford's School of Geography, and if geographers of both genders shared their discipline's plight of being in but not convincingly of Oxford, it is clear that women were in, rather than convincingly of, the School of Geography's staff until the end of the twentieth century. There were fewer of them; they were overrepresented in ancillary roles such as librarian and cartographer; they saw men appointed over them to prestigious University posts; they were even less likely than their male contemporaries to secure college tutorial fellowships; and they had to choose between a normal domestic life and a geographical career (all save one were childless and most were single). Noting the continued underrepresentation of women in the higher ranks of academe, McDowell remarks that 'the epistemological challenges of feminism may have changed ways of thinking, but many of the practical demands of the women's movement [for example, for childcare] have yet to be met' (2003, 619). Lest, however, this memoir end by casting women as victims, it is worth noting two final ways in which women were triumphantly in but not of Oxford's School of Geography. First, the school was the University institution best represented among militant suffrage activists, with the MacMunn sisters, Wharton and Fripp being vipers in the bosom of the school that anti-suffragist Mackinder had founded. Second, Marjorie Sweeting was a beacon of modernity in an otherwise rather complacently old-fashioned department. As Peter Hansen notes in his memoir of Kenneth Mason who, directly and through the tutors he appointed, shaped the school for much of the twentieth century, 'Mason's own publications, such as they were, were wholly devoted to surveying and mountaineering and, except for Marjorie Sweeting, the staff he appointed were products of the Oxford school with little commitment to research and publication. The school was thus ill-equipped to conduct research of the type that was beginning in both other Oxford schools and other British geography departments' (Hansen 2004). Sweeting was magnificently in but not of the school, a leading, and for a long time its sole, participant of the vibrant, collaborative and internationally recognized research culture that the school's head from the late 1970s had struggled to gear it towards.

In Oxford's School of Geography and the Environment, successor department to the School of Geography, past male geographers continue to dominate the daily life of present geographers. Students compete for the H. O. Beckit and A. J. Herbertson prizes and attend lectures and seminars in the Mackinder, Herbertson, Beckit or Gottmann rooms: Marjorie Sweeting is commemorated in the departmental tea room, something that she would have approved, being all in favour of collective food and drink as a way of promoting a pleasant collaborative spirit, but something which nonetheless perpetuates the dispiriting confinement of women to the domestic arena while their male colleagues dominate the academic spaces. We hope this chapter goes some way to secure women's place in Oxford geography's past and point to ways in which their future can be fostered.

We must conclude briefly by considering the question of whether the women recorded here were 'in but not of geography', that is, bodily within a thing called a 'School of Geography', but working within an understanding of 'geography' framed such to make them and other women and their work incidental to the discipline's history and canon (see, for example, Maddrell 2009; and relatedly the debate over what constitutes geography between Johnston (2005; 2007) and Driver and Baigent (2007)). The nineteenth- and early-twentieth-century stereotypes of male explorers on the one hand and women travellers or even 'female globetrotteresses' on the other were drawn precisely to render the same action geographical when done by men but not when done by women (Bell and McEwan 1996). Such distinctions were so crass that their advocates are now exposed to the same ridicule that they intended to heap on women, but this does not mean that 'geography' in all its aspects is now understood gender neutrally: Maddrell for example (2009, ch.3) neatly characterizes the geographical writing, compiling and editing that significant numbers of women undertook as 'liminal' geographical work: Were those who did it hired helps or disciplinary gatekeepers? Is our categorization of geographical work cause, consequence or both of the gender of those doing the work? The exploration of such questions at Oxford, Britain's first and one of its biggest geography departments, takes on a more than local importance, and we hope that the answers that this essay has offered form an effective challenge to our discipline's canonical history.

Acknowledgements

We should like to thank the very many archivists, librarians and others, and their institutions who have helped us write this article. We hope they enjoy reading it. (All institutions are at Oxford unless otherwise stated.) Elaine Archbold, Special Collections, University of Newcastle Library; Hugh Baldwin, National Oceanographic Library, University of Southampton; James Fishwick, librarian, Lady Margaret Hall; Gigi Horsfield, Radcliffe Science Library; Janet Howarth; Dr Alexander du Toit, Archive Cataloguer, Goldsmiths' College, University of London; Somerville College; the Bodleian Library; University of Oxford Archives; University of Sydney Archives; the archivists of St Hugh's and St Anne's Colleges, and Cheltenham Ladies' College; Jessica Osborne, *Daily Information*; SoGE Alumni and Development team; SoGE Communications team; Oxfordshire Centre for Local History; various past and present students and staff of the SoGE, including, but not limited to, those who are named as sources for distinct pieces of information.

Notes

1 Scargill's statement does not accord with the evidence of the Accounts book of the School of the Geography for 1899–1910 which suggests that women paid more than men for the diploma course but this remains to be investigated more fully.

2 Oxford's first diploma, in education, took its first students in 1897 and, of the 154 diplomas awarded between 1903 and 1911, 91 went to women (Honey and Curthoys 2000, 556); instruction for the diploma in education could also be had in the School of Geography (*The Geographical Teacher* 1,2 1902, 90).

3 Maddrell (2009, 155) suggests that H. O. Beckit's wife Elizabeth/Bessie was also 'part of the geographical community' in Oxford. We have traced her in a field trip photograph of 1913 (typescript dissertation listing in Radcliffe Science Library, Bodleian Libraries, Oxford) but her role on the trip and in the school more generally is not yet clear and awaits fuller investigation.

4 This is not named after Doreen Massey.

References

Primary sources

Church of the Province of South Africa, Parish Registers, 1801–2004.

Death certificate (Fanny Louisa Dorothy Herbertson).

Goldsmiths' College, Staff Registers, 1905–1936. Mss Archives, Goldsmiths' College, University of London.

Guide to Geographical Books and Appliances: Hints to Teachers and Students on the Choice of Geographical Books for Reference and Reading. 2nd edn revised by A. J. Herbertson and J. F. Unstead and N.E. MacMunn. London: George Philip, 1910.

'Home Office, Registered Papers, Suffragettes Meetings, Outrages, etc.' The National Archives, Kew, London, HO 144/1193/220196.

Institute of Heraldic and Genealogical Studies, Andrews Collection, Newspaper Index Cards, 1790–1976.

Kennedy, B. A. (1997), Curriculum vitae. Mss Archives, St. Hugh's College, University of Oxford.

Lady Margaret Hall, Register. Mss Archives, Lady Margaret Hall, University of Oxford.

London, Non-conformist Registers, 1694–1931, Mss Archives, City of London Corporation Libraries, Archives and Guildhall Art Gallery Department.

Marriage certificate (Fanny Louisa Dorothy Richardson Herbertson).

Royal Canadian Navy, Admirals' Medal recipients, http://www.navy-marine.forces.gc.ca/en/about/in-depth-admirals-medal-recipients.page

School of Geography, Cash analysis books, 1899–1964. Mss Archives, Geography archives GE3, University of Oxford Archives.

School of Geography, Matriculation forms, 1932–2000. Mss Archives, School of Geography, University of Oxford.

Sharp (Thomas) Archive, 1932–1984. Mss Archives, University of Newcastle, https://archiveshub.jisc.ac.uk/data/gb186-ths/

South Africa, Natal Province, Deaths 1863–1955.

Teachers' Registration Council (1914–1948), Registers. Mss Archives, Society of Genealogists, London.

United Kingdom, Board of Trade: Commercial and Statistical Department, Inwards Passenger Lists, 1878–1960. Series BT26, The National Archive, Kew.

University of Oxford, Academic Administration Division, Student data management and analysis. 'Student statistics, 2017', https://www1.admin.ox.ac.uk/aad/studentregistry/sdma/statistics/student/ (accessed 3 July 2018).

University of Oxford, University Registry, Register for the Diploma and Certificates in Geography, 1901–1939. Mss Archives UR 3/5/33/1–2, University of Oxford.

Women's Social and Political Union, 7th Annual Report of WSPU (1913), Home Office Registered Papers HO 45/10700/236973, The National Archive, Kew.

Secondary sources

A. J. R., ed. (1913), *The Suffrage Annual and Women's Who's Who*. London: Stanley Paul & Co.

Anon. (1889), 'Review of events: Presentation day at the University of London', *The Englishwoman's Review of Social and Industrial Questions* 20, 259–62.

Anon. (1897), 'University of Wales: The Gilchrist Scholarship', *South Wales Daily News*, 23 October 1897, no. 7909, p. 4, col. 5.

Anon. (1899), [Death notice for J. A. McMunn], *British Medical Journal*, 23 December 1899, 1770.

Anon. (1902), 'University and educational intelligence', *The Englishwoman's Review of Social and Industrial Questions* 33(255), 257–66.

Anon. (1904), 'Local intelligence', *Alloa Advertiser*, 23 April 1904, p. 2, col. 6.

Anon. (1907), 'Review: The Victoria History of the County of Essex, Volume 2', *Essex Review* 16, 101–8, https://archive.org/stream/essexreviewanil00fellgoog/essexreviewanil00fellgoog_djvu.txt (consulted July 2018).

Anon. (1914), 'Court circular: Social and personal', *Newcastle Journal*, 30 November 1914, p. 4, col. 5.

Anon. (1916), [Obit for Fanny Herbertson], *Geographical Teacher* 8, 211.

Anon. (1939), 'Sir W. Mitchell Ramsay [Obituary]', *Birmingham Mail*, 22 April 1939, p. 8, col. 3.

Anon. (1999), 'Moira Dunbar (Obituary]', *The Herald (Glasgow)*, 11 December 1999, 18.

Anon. (2014), 'Barbara Kennedy', obituary, *St Hugh's College Chronicle* 2013–14, 61–3.

Arthur, P. L. (2014), 'Biographical dictionaries in the digital era', in K. Bode and P. L. Arthur (eds), *Advancing Digital Humanities: Research, Methods, Theories*. London: Palgrave Macmillan, 83–94.

Arthur, P. L. (2015), 'Re-imagining a nation: The *Australian Dictionary of Biography* online', *European Journal of Life Writing* 4, 108–24.

Association for Promoting the Education of Women in Oxford. (1895), *Admission of Women to the B.A. Degree at Oxford: Report on Evidence Contained in Documents, or in Answers to Circular and Other Letters*. Oxford: The Council of the Association.

Auchmuty, R. (2008), 'Early women law students at Cambridge and Oxford', *Journal of Legal History* 29, 63–97.

Baigent, E. (2004a), 'Herbertson, Andrew John (1865–1915)', in H. C. G. Matthew and B. H. Harrison (eds), *Oxford Dictionary of National Biography*. Oxford: Oxford University Press.

Baigent, E. (2004b), 'Founders of the Royal Geographical Society of London (act. 1828–1830)', in H. C. G. Matthew and B. H. Harrison (eds), *Oxford Dictionary of National Biography*. Oxford: Oxford University Press.

Beckit, H. O. (1915), 'Andrew John Herbertson, 1865–1915', *Geographical Journal* 46, 319–20.

Bell, M. and McEwan, C. (1996), 'The admission of women fellows to the Royal Geographical Society, 1892–1914: The controversy and the outcome', *Geographical Journal* 162, 295–312.

Bentivolgio, P. and Doolan, S. (2014), *The History of a Bentivoglio Family in Australia*. Yamba: Bentivoglio.

Bradley, K. (1997), 'Faith, perserverance and patience: The history of the Oxford suffrage and anti-suffrage movements, 1870–1930'. unpublished PhD dissertation, Oxford Brookes University.

Brock, M. G. and Curthoys, M. C., eds (1997 and 2000), *The History of the University of Oxford: Volume VII: Nineteenth-Century Oxford*, 2 parts. Oxford: Clarendon Press.

Bromley, J. and Bromley, D. (2015), *Wellington's Men Remembered Volume 2: A Register of Memorials to Soldiers Who Served in the Peninsula War and at Waterloo* 1808–1815. Barnsley: Praetorian Press.

Brown, R. N. R. (1948), 'Scotland and some trends in geography: John Murray, Patrick Geddes, and Andrew Herbertson', *Geography* 33, 107–20 (Herbertson memorial lecture).

Butlin, R. A. (2004), 'Lucas, Sir Charles Prestwood (1853–1931)', in H. C. G. Matthew and B. H. Harrison (eds), *Oxford Dictionary of National Biography*. Oxford: Oxford University Press.

Buxton, E. (1984), 'Mary Marshall (1917–1983)', *Brown Book* (LMH college magazine) 1984, 35–6.

Carter, C. C. and McGregor, C. (1901–1902), 'Long vacation course at the School of Geography, Oxford', *Geographical Teacher* 1, 172–9.

Church, M. (2014), 'Deconstructing geomorphology: An appreciation of the contributions of Barbara A. Kennedy (1943–2014)', *Earth Surface Processes and Landforms* 39, 1269–72.

Clout, H. (2003), 'Place description, regional geography and area studies: The chorographic inheritance', in R. Johnston and M. Williams (eds), *A Century of British Geography*. Oxford: Oxford University Press for the British Academy, 247–74.

Clout, H. and Gosme, C. (2003), 'The Naval Intelligence Handbooks: A monument in geographical writing', *Progress in Human Geography* 27, 153–73.

Coones, P. (1989), 'The centenary of the Mackinder Readership at Oxford', *Geographical Journal* 155, 13–23.

Cossar, J. (1915), 'Obituary: Professor A. J. Herbertson', *Scottish Geographical Magazine* 31, 486–90.

Crawford, O. G. S. (1955), *Said and Done: The Autobiography of an Archaeologist*. London: Weidenfeld and Nicolson.

Crawford, E. (2006), *The Women's Suffrage Movement in Britain and Ireland: A Regional Survey*. Abingdon: Routledge.

Creagh, O'M. and Humphris, E. M. (1924), *The V. C. and D. S. O.*: A Complete Record *of All Those Officers, Non-Commissioned Officers and Men of His Majesty's Naval, Military and Air Forces Who have been Awarded these Decorations*, 3 vols. London: The Standard Art Book Company.

Currie, R. (1994), 'The arts and social studies, 1914–1939', in B. H. Harrison (ed.), *The History of the University of Oxford Volume VIII: The Twentieth Century*. Oxford: Clarendon Press, 109–38.

Denecke, H (1959), 'Obituary of Agnes Margaret Ramsay', *Brown Book* (LMH college magazine), 1959.

Driver, F. (2013), 'Hidden histories made visible? Reflections on a geographical exhibition', *Transactions of the Institute of British Geographers* 38, 420–35.

Driver, F. and Baigent, E. (2007), 'Biography and the history of geography: A response to Ron Johnston', *Progress in Human Geography* 31, 101–6.

Ellis, S., ed. (2013), *Eighty Years of Geography at Hull: A History of the Department of Geography University of Hull 1928–2008*. Hull: University of Hull, Department of Geography.

Firth, C. H. (1918), *The Oxford School of Geography*. Oxford: Blackwell.

Fleure, H. J. (1965), 'Recollections of A. J. Herbertson', *Geography* 50, 348–9.

Ford, D. (2016), 'The science of caves and karst: From the beginning of the Geological Society of America to ca. 1960', in J. M. Feinberg, Y. Gao and E. C. Alexander, Jr (eds), *Caves and Karst Across Time* (Geological Society of America Special Papers, 516). Boulder: Geological Society of America, 1–17.

Fox, H. (2004), 'Williams, Ivy (1877–1966)', in H. C. G. Matthew and B. H. Harrison (eds), *Oxford Dictionary of National Biography*. Oxford: Oxford University Press.

Frances, H. (1998), '"Pay the Piper, Call the Tune!": The Women's Tax Resistance League', in M. Joannou and J. Purvis (eds), *The Women's Suffrage Movement: New Feminist Perspectives*. Manchester: Manchester University Press, 65–76.

Fraser, R. (2014), 'Educational books', in W. R. Louis (ed.), *The History of Oxford University Press Volume 3: 1896 to 1970*. Oxford: Oxford University Press, 442–69.

Fritillary. (Oxford inter-collegiate women's magazine) 1894–.

Geldart, W. M., et al. (1912), 'Oxford Home Students', *The Spectator* 108(4381) (15 June 1912), 949.

The Geographical Teacher. (1901–1926). London: Geographical Association.

Geography: Journal of the Geographical Association. (1927–). Sheffield: Geographical Association.

Gilbert, E. W. (1965a), 'Andrew John Herbertson 1865–1915: An appreciation of his life and work', *Geography* 50, 313–31.

Gilbert, E. W. (1965b), 'Andrew John Herbertson, 1865–1915', *Geographical Journal* 131, 516–19.

Gilbert, E. W. (1972), *British Pioneers in Geography*. Newton Abbot: David & Charles.

Goldman, L. (1995), *Dons and Workers: Oxford and Adult Education since 1850*. Oxford: Clarendon Press.

Goudie, A. S. (1995), 'Obituaries: Marjorie Sweeting 1920–1994', *Geographical Journal* 161, 239.

Goudie, A. S. (1998), 'Kenneth J. Mason, 1887–1976', *Geographers Biobibliographical Studies* 18, 67–72.

Goudie, A., Ford, D. and Sugden, D. (1987), 'Dr Marjorie Sweeting', *Earth Surface Processes and Landforms* 12, 445–8.

Griffin, P., ed. (1986), *St Hugh's: One Hundred Years of Women's Education in Oxford*. Basingstoke: Macmillan.

Halsey, A. H. (1994), 'Oxford and the British Universities', in B. H. Harrison (ed.), *The History of the University of Oxford Volume VIII: The Twentieth Century*. Oxford: Clarendon Press, 577–606.

Hansen, P. H. (2004), 'Mason, Kenneth, (1887–1976)', in H. C. G. Matthew and B. H. Harrison (eds), *Oxford Dictionary of National Biography*. Oxford: Oxford University Press.

Harrison, B. H. (1978), *Separate Spheres: The Opposition to Women's Suffrage in Britain*. Abingdon: Routledge.

Harrison, B. H., ed. (1994), *The History of the University of Oxford* Volume VIII: *The Twentieth Century*. Oxford: Clarendon Press.

Hart, J. K. (2007), 'The role of women in British Quaternary science', in C. V. Burek and B. Higgs (eds), *The Role of Women in the History of Geology*. (Geological Society of London Special Publication 281) London: Geological Society of London, 83–95.

Hawthorn, G. (1965), 'Geography', *Oxford Magazine*, 4 March 1965, 256–7.

Herbertson, A. J. (1902), 'Geography in the university', *Scottish Geographical Magazine* 18, 124–132.

Heyes, J. F. (1886), 'A neglected subject', *Oxford Magazine*, 27 January 1886, reprinted as a pamphlet.

Heyes, J. F. (1887), 'The beginnings of geography in Oxford', *Oxford Review*, 2 November 1887, reprinted as a pamphlet.

Honey, J. R. De S. and Curthoys, M. C. (2000), 'Oxford and schooling', in M. G. Brock and M. C. Curthoys (eds), *The History of the University of Oxford: Volume VII: Nineteenth-Century Oxford, Part 2*. Oxford: Clarendon Press, 543–69.

Howarth, O. J. R. (1951), 'Centenary of section E (geography) in the British Association', *Scottish Geographical Magazine* 67, 145–60.

Howarth, J. (1994), 'Women', in B. H. Harrison (ed.), *The History of the University of Oxford Volume VIII: The Twentieth Century*. Oxford: Clarendon Press, 345–75.

Howarth, J. (2000a), 'The self-governing university, 1882–1914', in M. G. Brock and M. C. Curthoys (eds), *The History of the University of Oxford: Volume VII: Nineteenth-Century Oxford, Part 2*. Oxford: Clarendon Press, 599–643.

Howarth, J. (2000b), 'In Oxford but…not of Oxford': The women's colleges', in M. G. Brock and M. C. Curthoys (eds), *The History of the University of Oxford: Volume VII: Nineteenth-Century Oxford, Part 2*. Oxford: Clarendon Press, 236–307.

Hulbe, C. L., Wang, W. and Ommanney, S. (2010), 'Women in glaciology, a historical perspective', *Journal of Glaciology* 56, 944–64.

Institute of British Geographers, *Transactions*, 1935–1939.

Jay, L. J. (1965a), 'A. J. Herbertson: his services to school geography', *Geography* 50, 350–61.

Jay, L. J. (1965b), 'The published works of A. J. Herbertson: a classified list', *Geography* 50, 364–70.

Jay, L. J. (1979), 'Andrew John Herbertson, 1865–1915', *Geographers: Biobibliographical Studies* 3, 85–92.

Johnston, R. (2005), 'Learning our history from our pioneers: UK academic geographers in the Oxford Dictionary of National Biography', *Progress in Human Geography* 29, 651–67.

Johnston, R. (2007), 'Whose biography; whose history? a response to Driver and Baigent', *Progress in Human Geography* 31, 107–9.

Kay, J. (2007), 'No time for recreations till the vote is won'? Suffrage activists and leisure in Edwardian Britain', *Women's History Review* 16, 535–53.

Kearns, G. (1985), 'Halford John Mackinder, 1861–1947', *Geographers: Biobibliographical Studies* 9, 71–86.

Kearns, G. (2009), *Geopolitics and Empire: The Legacy of Halford Mackinder* (Oxford Geographical and Environmental Studies Series). Oxford: Oxford University Press.

Kearns, G. (2010), 'Geography, geopolitics and empire', *Transactions of the Institute of British Geographers* 35, 187–203.

Keighren, I. M. (2010), *Bringing Geography to Book: Ellen Semple and the Reception of Geographical Knowledge*. London: IB Tauris

Kennedy, B. A. (1995), 'Obituaries: Marjorie Sweeting', *Independent*, 18 January 1995, 16.

Library Association Record (1899–2002). London: Library Association.

Liddington, J. and Crawford, E., eds (2014), *Vanishing for the Vote: Suffrage, Citizenship and the Battle for the Census*. Manchester: Manchester University Press.

Liddington, J., Crawford, E. and Maund, E. A. (2011), '"Women do not count, neither shall they be counted": Suffrage, citizenship and the battle for the 1911 census', *History Workshop Journal* 71, 98–127.

Livingstone, D. N. (1992), *The Geographical Tradition: Episodes in the History of a Contested Enterprise*. Oxford: Blackwell.

London Gazette. (1666–). London: The Stationery Office.

Lorimer, H. (2003), 'Telling small stories: Spaces of knowledge and the practice of geography', *Transactions of the Institute of British Geographers* 28, 197–217.

Louis, W. R., Gadd, I. A. and Eliot, S., eds (2013), *History of Oxford University Press*: Volume III*: 1896 to 1970*. Oxford: Oxford University Press.

Mackinder, H. J. (1931), 'The human habitat', (Presidential address to Section E (geography) delivered at the Centenary Meeting of the British Association, London, 18 September 1931) *Scottish Geographical Magazine* 47, 321–35.

Mackinder, H. J. and Sadler, M. E. (1891), *University Extension, Past, Present and Future*. London: Cassell.

Mackinder, H. J., MacMunn, N. E. and Elton, E. F. (1916), 'Obituary [A. J. Herbertson]', *Geographical Teacher* 8, 143–6.

MacMunn, G. F. (1930), *Behind the Scenes in* Many *Wars, Being the Military Reminiscences of Lieut.-General Sir George MacMunn*. London: John Murray.

Maddrell, A. (2008), 'The "Map Girls": British women geographers' war work, shifting gender boundaries and reflections on the history of geography', *Transactions of the Institute of British Geographers* 33, 127–48.

Maddrell, A. (2009), *Complex Locations: Women's Geographical Work in the UK 1850–1970*. Oxford: Wiley–Blackwell.

Maddrell, A. (2015), 'To read or not to read? The politics of overlooking gender in the geographical canon', *Journal of Historical Geography* 49, 31–8.

Marsden, W. E. (1988), 'Continuity and change in geography textbooks: Perspectives from the 1930s to the 1960s', *Geography* 73, 327–43.

Mason, K. (1938), 'Geography in Oxford', *Oxford Magazine*, 17 February 1938, 411–12.

Mayhall, L. E. N. (2003), *The Militant Suffrage Movement: Citizenship and Resistance in Britain, 1860–1930*. Oxford: Oxford University Press.

McDowell, L. (2003), 'Geographers and sexual difference', in R. Johnston and M. Williams (eds), *A Century of British Geography*. Oxford: Oxford University Press for the British Academy, 603–23.

Middleton, D. (2004), 'Taylor, Annie Royle (1855–1922) (including Taylor, Susette Martha (1860–1920))', in H. C. G. Matthew and B. H. Harrison (eds), *Oxford Dictionary of National Biography*. Oxford: Oxford University Press.

Murray, O. (2000), 'Ancient history, 1872–1914', in M. G. Brock and M. C. Curthoys (eds), *The History of the University of Oxford: Volume VII: Nineteenth-Century Oxford, Part 2*. Oxford: Clarendon Press, 333–60.

Myres, J. L. (1929), '"The Oxford Magazine" and geography: A retrospect over forty years', *Oxford Magazine* 47, 456–8.

Ockwell, A. and Pollins, H. (2000), 'Extension' in all its forms', in M. G. Brock and M. C. Curthoys (eds), *The History of the University of Oxford: Volume VII: Nineteenth-Century Oxford, Part 2*. Oxford: Clarendon Press, 661–88.

Ominde S. H. (1971), 'Biographical note [Samuel Baker]', in S. H. Ominde (ed.), *Studies in East African Geography and Development*. Berkeley and Los Angeles: University of California Press, xi–xii.

Orford, E. J. (1932), 'The "Teaching of Geography" exhibition', *Geography* 17, 216–18.

Oxford Mail (1928–). Oxford: Oxford Mail & Times.

Oxford Oratory, News (2009–). *Parish Priest's Report*. Oxford: Catholic Church of St. Aloysius Gonzaga, www.oxfordoratory.org.uk/downloads/ppreport09.pdf (consulted November 2018).

Oxford University Gazette (1870–). Oxford: Clarendon Press.

Park, J. (1988), 'The British suffrage activists of 1913: An analysis', *Past & Present* 120, 147–62.

Parry, R. (2006), 'Map libraries', in J. H. Bowman (ed.), *British Librarianship and Information Work* 1991–2000. Aldershot: Ashgate, 227–44.

Phillips, V. (2001), *Record of Award Holders in Science, Engineering and the Arts 1891 to 2000*. South Kensington, London: Royal Commission for the Exhibition of 1851.

Ploszajska, T. (1999), *Geographical Education, Empire and Citizenship: Geographical Teaching and Learning in English Schools, 1870–1944*. (Historical Geography Research Series 35) [London]: Institute of British Geographers, Historical Geography Research Group.

Powell, R. C. (2015a), 'Notes on a geographical canon? Measures, models and scholarly enterprise', *Journal of Historical Geography* 49, 2–8.

Powell, R. C. (2015b), 'Geography and politics redux? Introduction to the virtual issue', *Transactions of the Institute of British Geographers* 40, 596–602.

Prince, H. C. (2000), *Geographers Engaged in Historical Geography in British Higher Education 1931–1991*. (Historical Geography Research Series 36) London: Institute of British Geographers, Historical Geography Research Group.

Roche, J. (1994), 'The non-medical sciences, 1939–1970', in B. H. Harrison (ed.), *The History of the University of Oxford. Volume VIII: The Twentieth Century*. Oxford: Clarendon Press, 251–90.

Rogers, A. M. A. H. (1938), *Degrees by Degrees: The Story of the Admission of Oxford Women Students to Membership of the University*. Oxford: Oxford University Press.

Roots, I. (2004), 'Firth, Sir Charles Harding (1857–1936)', in H. C. G. Matthew and B. H. Harrison (eds), *Oxford Dictionary of National Biography*. Oxford: Oxford University Press.

Rowley, D. (2008), 'Moira Dunbar', *Canadian Encyclopedia*, https://www.thecanadianencyclopedia.ca/en/article/isobel-moira-dunbar

St Hugh's Chronicle (1928–).

Scargill, D. I. (1976), 'The RGS and the foundations of geography at Oxford', *Geographical Journal* 142, 438–61.

Scargill, D. I. (1999), 'The Oxford School of Geography', *School of Geography Research Papers*, 55.

[School of Geography, Oxford] 'Remembering Andrew John Herbertson (1865–1915)', (2015), http://www.geog.ox.ac.uk/news/articles/150731-remembering-herbertson.html (consulted 27 July 2018).

Schwartz, V. R. and Przyblyski, J. M., eds (2004), *The Nineteenth-Century Visual Culture Reader*. Abingdon: Routledge.

Sheppard, J. A. (1994), *A Century of Geography: A History of the Geography Department of East London College (to 1935), Queen Mary College (1935–1989), Queen Mary and Westfield College (1989–)*. London: Department of Geography, Queen Mary and Westfield College.

The Ship (the magazine of the Old Students' Association of the Oxford Society for Home Students) 1935–.

Smee, D. K. (1953), 'Miss Blanche Hosgood', *Geography: Journal of the Geographical Association* 38, 320.

Stamp, L. D., ed. (1937), *The Land of Britain: The Report of the Land Utilisation Survey of Britain*. London: Geographical Publications Ltd.

Steel, R. W. (1987), *British Geography 1918–1945*. Cambridge: Cambridge University Press.

The Suffragette (1912–1915). London: Women's Social and Political Union.

Symonds, R. (1986), *Oxford and Empire: The Last Lost Cause?* Oxford: Clarendon Press.

Symonds, R. (2000), 'Oxford and the Empire', in M. G. Brock and M. C. Curthoys (eds), *The History of the University of Oxford: Volume VII: Nineteenth-Century Oxford, Part 2*. Oxford: Clarendon Press, 659–716.

Thomas, C. (2000), 'Moira Dunbar: Woman scientist the navy refused to take on board' obituary', *Guardian*, 12 January 2000, p. 22, col. 4.

University of Oxford (1877–1968), *The Examination Statutes ... Together with the Decrees of Convocation and Regulations of the Boards of Studies ...* . Oxford: Clarendon Press.

Unstead, J. F. (1965), 'A. J. Herbertson and his work', *Geography* 50, 343–7.

Viles, H. A. (1996), 'Marjorie Sweeting, 1920–1994', *Transactions, Institute of British Geographers* 21, 429–32.

Viles, H. A. (1997), 'A lifetime of landforms: Marjorie Sweeting's work on tropical and subtropical karst', in P. W. Williams (ed.), *Tropical and Subtropical Karst: Essays Dedicated to the Memory of Dr Marjorie Sweeting.* (*Zeitschrift für Geomorphologie, Supplementband* 108) Stuttgart: Borntraeger, 1–4.

Vincent, M. and Simpson-Lewis, W. (2009), 'Northern Exposure', Royal Canadian Geographical Society, http://www.rcgs.org/about/history/northern_exposure.asp (consulted September 2018).

Votes for Women (1907–1918). London: The Reformer's Press.

Walford, R. (2001), *Geography in British Schools, 1850–2000: Making a World of Difference*. London: Woburn Press.

White, W. B. (2016), 'Science of caves and karst: A half century of progress', in J. M. Feinberg, Y. Gao and E. C. Alexander, Jr (eds), *Caves and Karst across Time*. (Geological Society of America Special Papers, 516) Boulder: Geological Society of America, 19–33.

Wiggins, S. (2009), 'Gendered spaces and political identity: Debating societies in English women's colleges, 1890–1914', *Women's History Review* 18, 737–52.

Wojtczak, H. (2002), *Women of Victorian Sussex: Their Status, Occupations, and Dealings with the Law 1830–1870*. Hastings: The Hastings Press, http://www.hastingspress.co.uk/history/darent.html (consulted July 2018).

Yuan Daoxian (1987), 'Karst geomorphologist of the decades', *Earth Surface Processes and Landforms* 12, 448–9.

Selected bibliography of geographical works by F. D. Herbertson

Herbertson also wrote, co-wrote, or edited several French, Greek and Latin language learning books.

1896 (as translator with P. de Rousiers as author). *The Labour Question in Britain*. London: Macmillan.

1899 (with A. J. Herbertson) *Man and His Work: An Introduction to Human Geography*. London: A. & C. Black (Translated into Czech 1906).

1901 *North America* (selected by F. D. Herbertson, edited with introduction by A. J. Herbertson). London: A. & C. Black (Descriptive geographies from original sources) [2nd edn., 1906; 3rd edn., 1919].

1902a *Africa* (selected by F. D. Herbertson, edited with introduction by A. J. Herbertson). London: A. & C. Black (Descriptive geographies from original sources).

1902b *Central and South America with the West Indies* (selected by F. D. Herbertson, edited with introduction by A. J. Herbertson). London: A. & C. Black (Descriptive geographies from original sources) [4th edn., 1928].
1903a *Asia* (selected by F. D. Herbertson, edited with introduction by A. J. Herbertson). London: A. & C. Black (Descriptive geographies from original sources) [2nd edn., 1907; 3rd edn., 1913].
1903b *Australia and Oceania* (selected by F. D. Herbertson, edited with introduction by A. J. Herbertson). London: A. & C. Black (Descriptive geographies from original sources).
1903c *Europe* (selected by F. D. Herbertson, edited with introduction by A. J. Herbertson). London: A. & C. Black (Descriptive geographies from original sources).
1906 *The British Empire*. London: A. & C. Black (Descriptive geographies from original sources) [F. D. Herbertson as sole author].
1907 (with A. J. Herbertson) *The Senior Geography*. (The Oxford Geographies) Oxford: Clarendon Press [5th rev. edn., 1918].
1909–12 *The Elementary Geography*, 7 vols. Oxford: Clarendon Press.
1911 (with A. J. Herbertson) *The World's Commercial Products*. (Harmsworth Business Library, vol. 3) London: Educational Book Co.
1912–13 *The Clarendon Geography*. Oxford: Clarendon Press (6 parts in 2 vols.) [2nd edn., 1918; rev. edn., 1930].
1920 'Leplay and social science', *Sociological Review* 12, 36–42.
1921 'Leplay and social science', *Sociological Review* 13, 46–8 and 108–10.
1946 Republished in whole (Victor Branford and Alexander Farquharson, eds), *Sociological Review* 38, 89–204.
1950 *The Life of Frédéric Le Play* (V. Branford and A. Farquaharson, eds). Ledbury, Herefordshire: Le Play House Press.
[As translator] Paul de Rousiers, *The Labour Question in Britain*. republished Ulan Press, 2012. London: Forgotten Books, 2017.
[selected by F. D. Herbertson; edited by A. J. Herbertson] *Asia* Series: Descriptive geographies from original sources 1907 | 2nd edn. |. London: A. and C. Black. (Classic Reprint), London: Forgotten Books, 2018.

Bibliography of works by Nora MacMunn

1906 'The areas of the orographical regions of England and Wales', *Geographical Journal* 27, 288–91.
1907 'Social and economic history', in W. Page and J. H. Round (eds), *The Victoria History of the County of Essex*, vol. 2. London: (Archibald Constable and Co.), 313–54.

1909 'Historical maps and atlases', *Geographical Teacher* 5, 35–7 and Historical Associations pamphlet 13.

1913 *The Upper Thames Country and the Severn–Avon Plain*. Oxford: Clarendon Press.

1914a 'Gibraltar', in A. J. Herbertson and O. J. R. Howarth (eds), *Oxford Survey of the British Empire*: vol. 1 *The British Isles*. Oxford: Clarendon Press, 511–16.

1914b 'Malta', in A. J. Herbertson and O. J. R. Howarth (eds), *Oxford Survey of the British Empire*: vol. 1 *The British Isles*. Oxford: Clarendon Press, 516–29.

1916 'The French and German Borderlands, II: Geographical Aspect of the Political History', *Geographical Teacher* 8, 312–16.

1922 (with G. Coster) *Europe: A Regional Geography*. Oxford: Clarendon Press (compiled from notes left by A. J. Herbertson).

Bibliography of works by Mary Marshall

1942/3 (with K. Mason, J. L. Myres and others) *Turkey*. Oxford: Oxford University Press for the Naval Intelligence Division (Geographical Handbook Series), 2 vols.

1943 (with L. D. Stamp, as series editor) *The Land of Britain: The Report of the Land Utilisation Survey of Britain* 56 *Oxfordshire*. London: Geographical Publications Ltd.

1946 (with H. Scott, K. Mason, and others) *Western Arabia and the Red Sea*. Oxford: Oxford University Press for the Naval Intelligence Division (Geographical Handbook Series).

Bibliography of works by Elspeth Buxton

1966 'Government and other official publications for social and economic geography', unpublished thesis for fellowship of the Library Association.

1993 (with C. Robinson) *Hayfield Road: Nine Hundred Years of an Oxford Neighbourhood*. Oxford: C. Robinson.

Selected bibliography of works by Dorothy Forsaith

1920 (with H. Brown Smith) 'Geography Syllabus for Standards I. and II: The foundations of geography', *Geographical Teacher* 10, 222–3.

1932 *A Handbook for Geography Teachers*. London: Methuen (Goldsmith's college handbooks for teachers). [6th enlarged edn 1974].

1935 *Other Children's Homes*. London: A. & C. Black (Black's Graded Geographies).

1936 'Geography for children under eleven: The daily and seasonal rhythm', *Child Life: A Monthly Magazine for Teachers of Boys and Girls of 5–11 years* New Series 2, 98–100.

1951 *Many People in Many Lands*. London: University of London Press (Fundamental Geography, Book 1) [3rd edn., 1958].

Bibliography of works by Marie Bentivoglio

1921 'Notes on cassiterite crystals from the New England district, New South Wales, and Stanthorpe, Queensland', *Journal and Proceedings of the Royal Society of New South Wales* 55, 65–82.

1922 (with J. Read and H. G. Smith) 'Piperitone. Part III: The oximes of dl-piperitone', *Journal of the Chemical Society, Transactions* 121, 582–93.

1927 'An investigation of the rate of growth of crystals in different directions', *Proceedings of the Royal Society of London. Series A, Containing Papers of a Mathematical and Physical Character* 115(770), 59–87.

1928 'Geography and the child', *Australian Geographer* 1, 75–6.

1931 (with F. F. Friederich) *A Physical and Practical Geography for Secondary Schools*. Sydney: Whitcombe & Tombs.

1945 (with B. E. Cash) 'Flame-resistant cellulose acetate moulding compositions: Properties', *Modern Plastics* 22, 102–4.

1948 'Practical discussion . . . of films for prepackaging', *Pre-Pack-Age*, February, 24–7.

Bibliography of works by Dorothy Doveton

1936 'The economic geography of Swaziland', *Geographical Journal* 88, 322–31.

1937 *The Human Geography of Swaziland*. (Transactions of the Institute of British Geographers 8) London: George Philip and Sons (special issue, Doveton as sole author).

Selected bibliography of geographical works by En-Lan Liu

1931 'Climate: A dictator of China', unpublished MA dissertation, Clark University, Mass.

1936 'Rainfall variability in China', *Acta Geographica Sinica* 3(3), 529–32. (In Chinese with English abstract)

1937 'Pootu: A lost island', *Economic Geography* 13(2), 132–8.

1940 'Rainfall variations in China: Their nature, causes and effects', unpublished DPhil dissertation, University of Oxford.

1945 'Forgotten lands of the Szechwan border', *Geographical Magazine* 17(10), 438–451.
1946a 'The physical setting of China', *Journal of Geography* 45(1), 1–9.
1946b 'Climate and China's agricultural industry', *Journal of Geography* 45(3), 90–6.
1946c 'Reconstruction and geography', *National Reconstruction* 7(October), 37–47.
1948a 'Report on the Gia-rung of Ssu-tu in China', *Eastern World* 2(1–2).
1948b 'Mountain communities in West Sichuan', *Acta Geographica Sinica* 14(suppl.), 27–9 (In Chinese with English abstract).
1952 'The Ho-Si corridor', *Economic Geography* 28(1), 51–6.

Selected bibliography of works by Moira Dunbar

1951 'High latitude navigational flights', *Arctic* Circular 4, 88–92.
1952 (with L. S. Koenig, K. R. Greenaway and G. Hattersley-Smith) 'Arctic ice islands', *Arctic* 5, 67–103.
1953 (with J. V. Helk) 'Ice islands, evidence from North Greenland', *Arctic* 6, 263–71.
1954 'The pattern of ice distribution in Canadian Arctic seas' (Abstract of paper presented at a Symposium on Hudson Bay, 2 June 1954 arranged by the Canadian Committee on Oceanography for the Oceangraphic Session of Section V of the Royal Society of Canada), *Proceedings and Transactions of the Royal Society of Canada* Series 3, 48, 50.
1956a (with K. R. Greenaway) *Arctic Canada from the air*. Ottawa: Canada Defence Research Board.
1956b 'Notes on sea ice observed from C. G. S. D'Iberville and C. D. Howe, 13 August–12 October 1955', *Arctic report / Defence Research Board Ottawa*, no. 4/55.
1958 'The Royal Arctic Theatre', *Canadian Art* 15, 110–13.
1960a 'Thrust structures in young sea ice', *Journal of Glaciology* 3, 724–32.
1960b 'Translation of text accompanying the maps in the Atlas of Heat Balance', *Publication in meteorology / Defence Research Board Ottawa* 19.
1961 [Book review] 'La France et l'exploration polaire: de Verrazano a La Perouse, 1523–1788, by Marthe Emmanuel', *Arctic* 14, 129–31.
1962a 'The drift of North Pole 7 after its abandonment', *Canadian Geographer* 6, 129–42.
1962b 'Note on the formation process of thrust structures in young sea ice', *Journal of Glaciology* 4, 147–50.
1962c 'Ice conditions in the areas visited by C. M. S. John A. Macdonald in the summer of 1961', Canada Defence Research Board, *Directorate of Physical Research (G)* (Series) 9.
1963 (with W. W. Wittman) 'Some features of ice movement in the Arctic Basin', in *Proceedings of the Arctic Basin Symposium, October 1962* (Held at Hershey, Pa.,

under the auspices of the Arctic Institute of North America and the Office of Naval Research, US Navy). Washington DC: Arctic Institute of North America, 90–108.

1964 'Geographic distribution of superstructure icing in the northern hemisphere', Defence Research Board Ottawa, Directorate of Physical Research.

1965a 'Canadian proposal for changes in WMO sea ice terminology', *Polar Record* 12, 717–22.

1965b [Book review] 'Russian Settlement in North, by Terence Armstrong', *Arctic* 18, 280–2.

1967 'Discussion of Tsurikov, V. L. international ice nomenclature', *Oceanology USSR* 7(6), 879–82. (Also published in Russian original: Okeanologiîa 7, 1128–31.)

1969a 'The geographical position of the North Water', *Arctic* 22, 438–41.

1969b [Translation] 'A study of ice dynamics in the Gulf of Tartary by radio–physical methods', (Issledovaniîa dinamiki l⊠dov Tatarskogo proliva radiofizicheskimi metodami / Zhmurko, V. ÎA.) Translated from Trudy Arkticheskovo i Antarkticheskovo Nauchno-issledovatel⊠skogo Instituta, 284, 70–5, (1968) Ottawa: Defence Scientific Information Service, Defence Research Board, T 526 R.

1971a 'Anton and Dmitriy: Two Russian members of Scott's last expedition', *Polar Record* 15, 499–502.

1971b 'Winter ice reconnaissance in Nares Strait 1970–71', *Technical note / Defence Research Establishment Ottawa*, no. 71–34.

1972a 'Increasing severity of ice conditions in Baffin Bay and Davis Strait and its effect on the extreme limits of ice', in T. Karlsson (ed.), *Sea ice (Proceedings of an International Conference, Reykjavik, Iceland 10–13 May 1971)*. Reykjavik: National Research Council of Iceland, 87–93.

1972b (with M. J. Dunbar) 'The history of the North Water', *Proceedings of the Royal Society of Edinburgh, Section B – Biology* 72, 231–41.

1973a 'Ice regime and ice transport in Nares Strait', *Arctic* 26, 282–91.

1973b 'Women in science: How much progress have we really made?', *Science Forum* 6(2), 13–15.

1973c 'Winter regime of North-Water', *Transactions of the Royal Society of Canada* 11, 276–81.

1974 'Winter ice reconnaissance in Nares Strait 1972–73', *Technical note / Defence Research Establishment Ottawa*, 73–26.

1975a 'Interpretation of SLAR imagery of sea ice in Nares Strait and the Arctic Ocean', *Journal of Glaciology* 15, 193–213.

1975b (with W. F. Weeks) 'Interpretation of young sea ice in the Gulf of St Lawrence using side-looking radar and infrared imagery', *Cold Regions Research and Engineering Laboratory: Research Report* 337.

1975c 'Interpretation of SLAR imagery of ice in Nares Strait and the Arctic Ocean', *Report / Defence Research Establishment Ottawa*, 712.

1978a 'Interpretation of ice imagery from original and modified versions of a real-aperture SLAR', *Report / Defence Research Establishment Ottawa* 770.

1978b 'Petermann Gletscher: Possible source of a tabular iceberg off the coast of Newfoundland', *Journal of Glaciology* 20, 595–7.

1979 'Fall ice drift in Nares Strait, as observed by sideways-looking airborne radar', *Arctic* 32, 283–307.

1980 (with J. E. Keys) *Robeson Channel ice drift and oceanographic measurements 1970–1975*. Ottawa: Defense Research Establishment.

1986 'Man and ice: Experience with the frozen ocean', in S. Milligan and W. O. Kupsch (eds), *Living explorers of the Canadian Arctic: The Historic Symposium of Arctic Scientists, Explorers, and Adventurers*, Toronto, 1978. Yellowknife: Outcrop, The Northern Publishers, 168–82.

Bibliography of works by Barbara Kennedy

1967a 'A comparative study of multiply-controlled valley asymmetry in S.E. Wyoming and S.W. Manitoba', unpublished MSc dissertation, University of British Columbia.

1967b (with M. A. Melton) *Stream-valley Asymmetry in an Arctic–Subarctic Environment: Conditions Governing the Geomorphic Processes*. (Research Papers, AINA–RP–42) Montreal: Arctic Institute of North America.

1969a 'Studies of erosional valley-side asymmetry', unpublished PhD dissertation, University of Cambridge.

1969b 'Periglacial morphometry', in R. J. Chorley (ed.), *Water, Earth and Man: A Synthesis of Hydrology, Geomorphology and Socio–Economic Geography*. London: Methuen, 381–8.

1971 (with R. J. Chorley) *Physical Geography: A Systems Approach*. London: Prentice-Hall International.

1972a '"Bankfull" discharge and meander forms', *Area* 4, 209–12.

1972b (with M. A. Melton) 'Valley asymmetry and slope forms of a permafrost area in the Northwest Territories, Canada', in R. J. Price and D. E. Sugden (eds), *Polar Geomorphology*. (Institute of British Geographers Special Publication 4) London: Institute of British Geographers, 107–21.

1976 'Valley-side slopes and climate', in E. Derbyshire (ed.), *Geomorphology and Climate*. Chichester: John Wiley & Sons, 171–201.

1977 'A question of scale?', *Progress in Physical Geography* 1, 154–7.

1978 'After Horton', *Earth Surface Processes* 3, 219–31.

1979 'A naughty world', *Transactions of the Institute of British Geographers* new series 4, 550–8.

1983 'On outrageous hypotheses in geography', *Geography* 68, 326–30.

1984 'On Playfair's law of accordant junctions', *Earth Surface Processes and Landforms* 9, 153–73.

1985a Contributions to A. S. Goudie, et al. (eds), *The Encyclopaedic Dictionary of Physical Geography*. Oxford: Basil Blackwell.

1985b Contributions to V. Fuchs (ed.), *Oxford Illustrated Encyclopedia*. Vol.1, *The Physical World*. Oxford: Oxford University Press.

1990 'Paint your boulder', in D. Barsch (ed.), *Geomorphology and Geoecology: General, Invited and Special Lectures (Proceedings of the 2*nd *International Conference on Geomorphology; 1989*, vol. 1*)*. (*Zeitschrift für Geomorphologie Supplementband* 79) Stuttgart: Borntraeger, 213–16.

1991 'The Darwin correspondence', [Guest editorial], *Journal of Biogeography* 18, 591–3.

1992a 'First catch your hare … Research designs for individual projects', in A. Rogers, H. Viles and A. Goudie (eds), *The Student's Companion to Geography*. Oxford: Basil Blackwell, 128–40.

1992b 'Hutton to Horton: Views of sequence, progression and equilibrium in geomorphology', *Geomorphology* 5, 231–50.

1993a '"… no prospect of an end"', *Geography* 78, 124–36.

1993b 'The end of equilibrium?', 3rd William Vaughan Lewis Seminar. Cambridge: Department of Geography, University of Cambridge, 1–12.

1994a 'Classics in physical geography revisited: Carter, C. S. and R. J. Chorley, 1961: Early slope development in an expanding stream system. *Geographical Magazine* 97, 117–30', *Progress in Physical Geography* 18, 397–400.

1994b 'Requiem for a dead concept', *Annals of the Association of American Geographers* 84, 702–5.

1997a 'The trouble with valleys', in D. R. Stoddart (ed.), *Process and Form in Geomorphology*. London: Routledge, 60–73.

1997b 'Classics in physical geography revisited: Schumm S. A. and R. W. Lichty, 1965. Time, space and causality in geomorphology in *American Journal of Science* 263: 110–119', *Progress in Physical Geography* 21, 419–23.

1999 'Flow and sedimentation at tributary river mouths: A comparison with mesotidal estuaries', in A. J. Miller and A. Gupta (eds), *Varieties of Fluvial Form*. Chichester: John Wiley & Sons, 409–26.

2000 'Trompe l'oeil', *Journal of Biogeography* 27, 37–8.

2001 'Charles Lyell and "Modern changes of the Earth": The Milledgeville Gully', *Geomorphology* 40, 91–8.

2004 'Classics in physical geography revisited: Stoddart D. R. 1966. Darwin's impact on geography in *Annals of the Association of American Geographers* 56: 683–698', *Progress in Physical Geography* 28, 399–403.

2008 *Inventing the Earth: Ideas on Landscape Development since 1740*. Oxford: Blackwell.

Bibliography of works by Marjorie Sweeting (excluding book reviews)

1943 'Wave-trough experiments on beach profiles', *Geographical Journal* 101, 163–8.

1947 (with V. Wilson and 17 others) 'The teaching of geology in schools: Report of a discussion held on 5th April, 1946', *Proceedings of the Geologists' Association* 58, 1–44.

1949 'The geomorphology of the carboniferous limestone of the Ingleborough District, North West Yorkshire', unpublished PhD dissertation, University of Cambridge.

1950 'Erosion cycles and limestone caverns in the Ingleborough district', *Geographical Journal* 115, 63–78.

1953 'The enclosed depression of Carran, Co. Clare', *Irish Geographer* 2, 218–24.

1955a 'The landforms of northwest County Clare, Ireland', *Transactions of the Institute of British Geographers* 21, 33–49.

1955b 'The karstlands of Jugoslavia', *Transactions of the Cave Research Group* 4, 55–72.

1955c 'Recent trends of glacier fluctuations in the north Atlantic area', *Nature* 175 (4448), 188–93.

1956a 'Hydrogeological observations in parts of the white limestone areas in Jamaica, B.W.I.', *Bulletin of the Geological Survey of Jamaica* 2, 1–22.

1956b (with G. Groom) 'Notes on the glacier fluctuations in Bünsow Land, Central Vestspitsbergen', *Journal of Glaciology* 2, 640–1.

1957a (with J. A. Steers and 10 others) 'The physique of Wales: Discussion', *Geographical Journal* 123, 221–30.

1957b 'Notes on the caves of Jamaica', *Transactions of the Cave Research Group* 5, 1–11.

1958a (with G. E. Groom) 'Valleys and raised beaches in Bünsow Land, Central Vestspitzbergen', *Norsk Polarinstitutt Skrifter* 115, 1–18.

1958b 'The karstlands of Jamaica', *Geographical Journal* 124, 184–99.

1958c (with W. W. Black) 'Easter field meeting in the Settle district: 8–11 April 1955', *Proceedings of the Geologists' Association* 69, 24–7.

1959a (with J. N. Jennings) 'Underground breach of a divide at Mole Creek, Tasmania' *Australian Journal of Science* 21, 261–2.

1959b (with W. V. Lewis) 'The karstlands of Jamaica: Cockpits or rounded hills?', *Geographical Journal* 125, 289–91.

1960a (with A. Gerstenhauer) 'Zu Frage der absoluten Geschwindigkeit der Kalkkorrosion in verschiedenen Klimaten', in H. Lehmann (ed.), *Internationale Beiträge zur Karstmorphologie*. (*Zeitschrift für Geomorphologie Supplementband* 2) Stuttgart: Borntraeger, 66–73.

1960b 'The caves of the Buchan area, Victoria, Australia', in H. Lehmann (ed.), *Internationale Beiträge zur Karstmorphologie*. (*Zeitschrift für Geomorphologie Supplementband* 2) Stuttgart: Borntraeger, 81–9.

1961 (with J. N. Jennings) 'Caliche pseudo-anticlines in the Fitzroy Basin, W. Australia', *American Journal of Science* 259, 635–9.

1962 (with J. A. Steers and 3 others) 'Cambridge expedition to British Honduras: Discussion', *Geographical Journal* 128, 171–3.

1963a (with J. N. Jennings) *The limestone Ranges of the Fitzroy Basin, Western Australia: A Tropical Semi-arid Karst*. (*Bonner Geographische Abhandlungen* 32) Bonn: In Kommission bei F. Dümmler.

1963b (with J. N. Jennings) 'The tunnel: A cave in the Napier Range, Fitzroy Basin, Western Australia', *Transactions Cave Research Group* 6, 53–68.

1964 'Some factors in the absolute denudation of limestone terrains', *Erdkunde* 18, 92–5.

1965a 'Denudation in limestone regions: A symposium: Introduction', *Geographical Journal* 131, 34–7.

1965b (with K. S. Sandford and 9 others) 'Denudation in limestone regions: A symposium: Discussion', *Geographical Journal* 131, 51–6.

1966a 'The weathering of limestones: With particular reference to the Carboniferous limestones of northern England', in G. H. Dury (ed.), *Essays in Geomorphology*. London: Heinemann, 177–210.

1966b (with J. N. Jennings) 'Old Napier Downs cave, West Kimberley, Western Australia', *Helictite* 4, 25–32.

1968a 'Some variations in the types of limestones and their relation to cave formation', *Proceedings 4th International Speleological Congress, Ljublijana, 1965*. Ljubljana: Fédération Yougoslave de Spéléologie, Section 3 (Physical Speleology), 227–32.

1968b 'The University of Edinburgh British Honduras–Yucatan Expedition Section C (Karst morphology)', *Geographical Journal* 134, 49–54.

1970a (with G. S. Sweeting) 'Some aspects of the carboniferous limestone in relation to its landform with particular reference to N. W. Yorkshire and County Clare', (Actes de la Réunion Internationale de Karstologie, Languedoc–Provence, 1968) *Etudes et travaux de Méditerranée* 7, 201–9.

1970b 'Recent developments and techniques in the study of karst landforms in the British Isles', *Geographia Polonica* 18, 227–41.

1971 'Caves of the Ingleborough Area, Yorkshire', *Geographical Journal* 137, 138–9.

1972a 'The karst of Great Britain', in M. Herak and V. T. Stringfield (eds), *Karst: Important Karst Regions of the Northern Hemisphere*. Amsterdam: Elsevier, 417–43.

1972b *Karst Landforms*. London: Macmillan.

1972c 'Karst and solution phenomena in the Santa Rosa area, New Mexico', in V. C. Keeley and F. D. Trauger (eds), *East-Central New Mexico: Guidebook [to the] Twenty Third Field Conference*. Socorro: New Mexico Geological Society, 168–70.

1972d (with G. S. Sweeting) 'Some aspects of the dolomites of the Guelph and Niagara areas, Ontario, Canada', in Comité nationale de géographie, Commission des Phenomenes Karstiques (ed.), *Actes du Colloque International de Karstologie et de Spéléologie (21–25 Aout 1971) (Languedoc, Rouergue, Quercy, Perigord)*. Le Mans: Centre Regional de Documentation de Caen, 85–95.

1973a 'Some problems of relict and fossil karst in Britain', *Geographische Zeitschrift* 32, 104–7.

1973b 'Some aspects of karst hydrology', *Geographical Journal* 139, 280–5.

1973c (with A. C. Waltham and 8 others) 'Some results and applications of karst hydrology: A symposium: Discussion', *Geographical Journal* 139, 305–10.

1973d 'Karst landforms and limestones', *South African Geographical Journal* 55, 81–8.

1973e (with M. E. Marker) 'Large-scale subsurface weathering features (pinnacles) in the Northern Cape Province, South Africa', in Comité nationale de géographie, Commission des Phenomenes Karstiques (ed.), *Actes du Colloque de Karstologie et de Spéléologie (24–29 Septembre 1973) (Pyrénées occidentales françaises et région cantabrique espagnole)*. Aix-en-Provence: Institut de Géographie publié avec le concours de l'E.R.A. 282 du C.N.R.S., 53–69.

1974a (with A. C. Waltham) (eds), *The Limestones and Caves of Northwest England*. Newton Abbot: David & Charles for The British Cave Research Association.

1974b 'Karst geomorphology in North-West England', in A. C. Waltham and M. M. Sweeting (eds), *The Limestones and Caves of Northwest England*. Newton Abbot: David & Charles for The British Cave Research Association, 46–78.

1974c 'The morphology of humid tropical karst', *Proceedings Geographical Association of Rhodesia* 6, 50–65.

1975a 'Present problems in the karst denudation research in the world', in I. Gams (ed.), *Karst Processes and Relevant Landforms: Proceedings of the International Symposium on Standardization of Field Research Methods of Karst Denudation (Corrosion)*. Ljubljana: University of Ljubljana, 7–15.

1975b 'Some comments on the lithological basis of karst landform variations', *Proceedings of the* 6th *International Congress of Speleology*, Olomouc, ČSSR *(1973)* vol. 1. Prague: Academia, 319–29.

1976a (with K. H. Pfeffer) (eds), *Karst Processes*. (*Zeitschrift für Geomorphologie: Supplementband* 26) Stuttgart: Borntraeger.

1976b 'Present problems in karst geomorphology', in M. M. Sweeting and K. H. Pfeffer (eds), *Karst Processes*. (*Zeitschrift für Geomorphologie: Supplementband* 26) Stuttgart: Borntraeger, 1–5.

1978a 'Some observations on New Zealand limestone areas', in J. L. Davies, M. A. J. Williams and J. N. Jennings (eds), *Landform Evolution in Australasia*. Canberra: Australian National University Press, 250–8.

1978b 'The landscape of one-seventh of China', *Geographical Magazine* 50, 393–400.

1978c 'Karst water and karst hydrology', *Progress in Physical Geography* 2, 99–106.

1978d 'The karst of Kweilin, southern China', *Geographical Journal* 144, 199–204.

1978e (with E. H. Brown and E. Derbyshire) 'British geographers in China, 1977', *Geographical Journal* 144, 187–207.

1979a 'Solution and erosion in the karst of Melinau limestone in Gunung Mulu National Park, Sarawak, Borneo', in G. Fabre and J. Nicod (eds), *Actes du symposium international sur l'erosion karstique*. Nîmes: Association Française de Karstologie, 61–5.

1979b 'Karst morphology and limestone petrology', *Progress in Physical Geography* 3, 102–10.

1979c 'Weathering and solution of the Melinau limestones in the Gunung Mulu National Park, Sarawak, Malaysia', *Annales de la Société Géologique de Belgique* 102, 53–7.

1979d (as editor) *Problems in Karst Environments*. (*Zeitschrift für Geomorphologie: Supplementband* 32) Stuttgart: Borntraeger.

1979e 'It always rains in Gunong Mulu', *Geographical Magazine* 51, 271–7.

1980a 'Karst and climate: Areview', in H. Hagedorn, M. Thomas and D. Busche (eds), *Perspectives in Geomorphology: Papers from the First British-German Symposium on Geomorphology*. (*Zeitschrift für Geomorphologie: Supplementband* 36). Stuttgart: Borntraeger, 203–16.

1980b 'Symposium on the geomorphology of the Mulu hills. 1. The geomorphology of Mulu: An introduction', *Geographical Journal* 146, 1–7.

1980c (with P. M. Thorpe and R. L. Otlet) 'Hydrological implications from 14C profiling of UK tufa', *Radiocarbon* 22, 897–908.

1981 (as editor) *Karst Geomorphology*. (*Benchmark Series in Geology* 59) Stroudsburg: Dowden, Hutchinson & Ross.

1982a (with N. Lancaster) 'Solutional and wind erosion forms on limestone in the Central Namib Desert', *Zeitschrift für Geomorphologie* Neue Folge 26, 197–207.

1982b (with H. A. Osmaston) 'Geomorphology [of the Gunung Mulu National Park, Sarawak]', *Sarawak Museum Journal*, 30(51), special issue 2, 75–93.

1982c 'Obituary: C. F. W. R. Gullick 1907–1981', *Transactions of the Institute of British Geographers* new series 7, 511–12.

1982d 'Meeting abstract: Karstlands of the world', *Journal of the Geological Society* 139, 220.

1983a (with M. E. Marker) 'Karst development on the Alexandria limestones E. Cape Province, South Africa', *Zeitschrift für Geomorphologie* Neue Folge 27, 21–38.

1983b 'The geological and morphological setting', in D. A. Roe (ed.), *Adlun in the Stone Age: The Excavations of D. A. E. Garrod in the Lebanon, 1958–1963*, (British Archaeological Report, *International Series*, 159) *Part 1*. Oxford: British Archaeological Reports, 11–21.

1983c (with M.J. Tooley and 6 others) 'Endangered coastlines of geomorphological importance: Discussion', *Geographical Journal* 149, 68–71.

1983d 'Editorial: Chinese karst', *Area* 15, 51.

1984 (with A. M. Rae and M. Ivanovich) 'Natural uranium in the cave of Son Muleta, Mallorca: A background study for uranium series dating of bones', in W. H. Waldren (ed.), *Early Settlement in the Western Mediterranean Islands and their Peripheral Areas*, (*British Archaeological Reports International Series* 229) *Part 2*. Oxford: British Archaeological Reports, 295–311.

1985a (with D. Yuan, M. Hu and P. A. Bull) 'Aspects of the Quaternary deposits around Guilin, Guangxi, South China. A scanning electron microscope study', *Carsologica Sinica* 3, 267–71.

1985b 'The work of Jean Corbel in the British Isles', *Spelunca mémoires* (Fédération française de spéléologie) 14, 68.

1986a (with K. Paterson) (eds), *New Directions in Karst: Proceedings of the Anglo–French Karst Symposium, September 1983*. Norwich: Geo Books.

1986b (with C. Ping and S. Hua) 'The pinnacle karst of the stone forst, Lunan, Yunnan, China: an example of subjacent karst', in K. Paterson and M. M. Sweeting (eds), *New Directions in Karst*. Norwich: Geo Books, 597–607.

1986c (with H. A. Brown and R. L. Otlet) 'Stable isotopes-an investigation into their application in karst hydrology in the UK with special reference to the Malham area, North Yorkshire', in K. Paterson and M. M. Sweeting (eds), *New Directions in Karst*. Norwich: Geo Books. 213–31.

1986d (with C. O. Hunt, S. J. Gale and K. Paterson) 'Palynology: a neglected tool in British cave studies', in K. Paterson and M. M. Sweeting (eds), *New Directions in Karst*. Norwich: Geo Books, 323–32.

1986e 'Limestone landscapes of South China', *Geology Today* 1, 11–15.

1988 (with H. Bao and D. Zhang) 'Karst in Tibet', in *Proceedings, 21st IAH Congress, Karst Hyydrology and Karst Environment Protection*, Guilin, China, 213–18.

1991 (with H. Bao and D. Zhang) 'The problem of palaeokarst in Tibet', *Geographical Journal*, 316–25.

1992 'Tectonics and fluvial denudation in the formation of cone karst, with particular reference to South China', *Tübinger Geographische Studien* 109, 45–56.

1993 'Reflections on the development of Karst geomorphology in Europe and a comparison with its development in China', *Zeitschrift für Geomorphologie* 37, 127–36.

1995 *The Karst of China: Its Geomorphology and Environment*, Marjorie M. Sweeting. Berlin, London: Springer, 1995.

Chronologies

Fanny Herbertson

1864	Born in Leamington, Warwickshire
1888	Graduated BA Classics from the University of London
1888–1893	Employed as Classics mistress at Cheltenham Ladies' College
1893	Married A. J. Herbertson in North Leith
1894	Moved to Manchester
	Son Andrew Hunter born
1896	Moved to Edinburgh
1899	Moved to Oxford: ran summer schools for teachers and wrote textbooks with Herbertson
	Daughter Margaret Alice Louise born
1909–12	Published bestselling seven-volume school textbook *Elementary Geography*
1915	Died of heart failure in Oxford
1920	Acclaimed biography of Frédéric Le Play published posthumously

Nora MacMunn

1875	Born in Chelsea
1896	Moved to St Leonards, Sussex
1903	Took FHS examinations in modern history, University of Oxford
1904	Took diploma in geography
1906	Appointed Demonstrator in School of Geography
1908	Marched in London with Oxford University Women's Suffrage Society
1910	Published *Guide to Geographical Books and Appliances* (with A. J. Herbertson and F. J. Unstead)
	Joined Women's Tax Resistance League
1911	Joined Oxford branch of Women's Social and Political Union (WSPU)
1912	Served on Committee for Society for Home Students
1918	Appointed Senior Demonstrator in geography
1921	Graduated MA by decree
1922	Appointed Assistant Lecturer in geography
1924	Appointed Lecturer in regional geography
1933	Supervised her first graduate student
1935	Retired from posts at Society for Home Students and School of Geography, and moved to Biddesden, near Andover, Hampshire
1936	Became member of Institute of British Geographers
1967	Died at Eastbourne, East Sussex

Mary Marshall

1917	Born in Doncaster
1939	Graduated BA in geography from Lady Margaret Hall
	Appointed research scholar at Lady Margaret Hall
1939–41	Worked as Cartographer to the Oxford War Agricultural Committee
1941–46	Appointed Geographer to the Naval Intelligence Division of the Admiralty, working on Naval Intelligence Handbooks on *Turkey; Western Arabia and the Red Sea;* and *Palestine and Transjordan*
1943	Published Land Utilisation Survey report on Oxfordshire
	Appointed examiner on shortened wartime Honour School
1944	Appointed lecturer in geography at Lady Margaret Hall
1945	Appointed Departmental Lecturer at School of Geography
1948	Appointed Susette Taylor Fellow at LMH
1954	Became Senior Lecturer at School of Geography
1961	Lectureship at LMH ended. Elected to an honorary research fellowship at the college
1965	Elected fellow of St Cross College
1982	Took early retirement owing to poor health
1983	Died in Oxford

Marjorie Sweeting

1920	Born in Fulham, London
1941	Graduated BA in geography from University of Cambridge
1942–45	Taught geography at Howell's School, Denbigh, Wales
1945–48	Completed a PhD at the University of Cambridge
1948–51	Research fellow at Newnham College, Cambridge
1951	Appointed Lecturer and Director of Studies in geography at St Hugh's College, Oxford
1952	Appointed University Lecturer in geography
1955	Received Royal Geographical Society's Gill Memorial Prize
1959	Awarded Certificate of Merit of the National Speleological Society of America
1969	Received Honorary Membership of the Cave Research Foundation of America
1972	Published *Karst Landforms*
1973	Medallist of Palacky University, Czechoslovakia
1973–75	Became first woman chair of the British Geomorphology Research Group
1976	Made Honorary Fellow of the National Speleological Society of America

1977	Appointed Reader in geography
	First visited China, when western geoscientists were admitted for the first time since the communist revolution
1980	Awarded Royal Geographical Society's Busk Medal
1983	Served as acting head of the School of Geography
1987	Retired
1994	Died of cancer in Oxford
1995	*The Karst of China* published

Barbara Kennedy

1943	Born in Ottawa, Canada
1945	Moved to the UK
1965	Graduated BA in geography from University of Cambridge
1965–67	Completed an MA at the University of British Columbia, Canada
1969	Graduated PhD from University of Cambridge
1968–72	Worked as Lecturer and Fellow at New Hall, Cambridge and Demonstrator in Department of Geography, University of Cambridge
1971	Published *Physical Geography: A Systems Approach* (with Richard Chorley)
1973–79	Lecturer at University of Manchester
1979	Appointed Lecturer in Geography at the Oxford School of Geography and Fellow and Tutor at St Hugh's College
1996–2000	Appointed Tutor for Admissions and then Senior Tutor at St Hugh's College
2000	Became University's International Recruitment Coordinator
2005	Retired, fellow emerita at St Hugh's
2008	Published *Inventing the Earth: Ideas on Landscape Development since 1740*
2014	Died of cancer in Oxford

3

Alice Louise Saunier-Seïté, 1925–2003

Hugh Clout

Alice Saunier (later Saunier-Seïté) was born into a humble family, and rose to be a minister in government. After training as a teacher, she took a degree in geography, which was followed by a state doctorate. She lectured for a time but soon developed a career in educational administration, as director of several new institutions and then as rector with responsibility for all aspects of state education in a French region. She was the first woman to hold such a position. Appointed secretary of state for universities in 1976, Madame Saunier-Seïté became minister for higher education two years later, being required to deal with the outcomes of revisions to teaching, learning and governance in state universities following legislation in the wake of social disturbances during 1968. Following the victory of a Socialist presidential candidate in 1981, she left national government and was elected professor of geography and planning at the Conservatoire National des Arts et Métiers, remaining there until retirement. She was a member of the Paris city Council and was involved with the management of many political and learned societies. In 1995 Alice Saunier-Seïté was honoured by election to membership of the Académie des sciences morales et politiques of the Institut de France. Her geographical expertise was in human adaptation to harsh environments, especially in Austria, Iceland, Scandinavia and Canada; many of her publications relate to these regions. She also wrote a number of historical biographies. Twice married, she had two sons with her first husband, Elie-Jacques Picard (1920–2007), a mathematician.

Education, life and work

Alice Louise Saunier was born on 26 April 1925 at Saint-Jean-le-Centenier, a village of 600 inhabitants in the southern part of Ardèche *département* in the Massif Central (Joly 2013, 93). Her parents, Daniel Saunier and Marie-Louise (née Lascombe), ran the local bakery. Daniel was a firm believer in the importance of education at a time when the village school was a respected institution in France that offered the chance of social advancement to bright children who worked hard. More than half a century later, his daughter noted that 'for him, it was the school, rather than universal suffrage or the courts, that guaranteed justice' in society (Saunier-Seïté 1982, 16). She was a quick learner and the headmaster advised her parents that she should be encouraged to take the entrance examination for secondary education at boarding school. Saunier-Seïté recalled that success followed 'several weeks of intensely hard work that landed me a grant. In September [1937], I left the village for the college (*école primaire supérieure*) at Privas where I spent the next four years . . . I only went home three times a year, covering the 30 km over the Coiron plateau on an old bicycle' (17). Discipline was strict in this 'secular convent' where many girls belonged to a variety of Huguenot (Protestant) sects, so that 'on Sunday mornings eight or ten different groups formed up in the playground' to walk to worship in various chapels (17). Saunier-Seïté remarked that 'this religious mosaic instilled tolerance in us all' (17).

Distractions from school work were few apart from supervised walks in the picturesque country lanes around Privas, when the girls had an opportunity to chat among themselves. The obligation to help with domestic chores repressed any 'blue stocking tendencies' (17). Four years were spent in this way, enabling Alice to acquire a good general education, including a knowledge of German that later proved very useful. The last two years were a particularly anxious time for her because of the material hardships of wartime, including lack of food and warm clothing. Even worse was the fact that the scholarship grant did not cover all expenses and her parents were scarcely able to support her financially as their business declined. With strong encouragement from the headmistress, who recognized her potential, Alice studied for the entrance examination for a teacher training college (*école normale*), which offered an escape route from her impoverishment. Successful students from such institutions were funded for three years to study pedagogy as well as preparing for the *baccalauréat* (school leaving) examination. On graduation they were sent to work for ten years in state schools; as civil servants they enjoyed a measure of job protection.

With a commitment to hard work as well as a strong will and independent spirit, Alice Saunier continued to enhance her studies. In the autumn of 1941, at the age of sixteen, she joined the teacher training college at Tournon, which was based in the town's old high school (*lycée*). Material conditions remained harsh. Half a century later, she remembered that

> the dormitory was glacial. We had to break off stalactites from dripping taps; the showers no longer worked. The boarders' fingers, toes and knees were swollen and black with chilblains. We shivered at night and in class. We were very, very cold … and the refectory was worse. We were very, very hungry. (cited in Mazeaud 2007)

Young German soldiers were billeted in part of the college. Alice 'felt pity rather than hatred toward them', but that did not prevent her taking messages to the Resistance in the city of Lyon and in the Vercors mountains (Mazeaud 2007; Choffel 1979, 12). Later, she married Elie-Jacques Picard, brother of one of her fellow students, who was a distinguished member of the Resistance.

Alice Saunier completed her training with a further year spent at the Teachers' College (*institut pédagogique*) at Valence, from which most graduates were sent to teach in single-class schools in villages in and around the Rhône valley.

In this way, the training colleges 'favoured and reinforced the vocation of school teaching and gave France its teachers who spread literacy [and basic education] across our country' (Saunier-Seïté 1982, 20). However, Saunier was frustrated by all this preparation and declared that her earlier studies had 'undermined the initial vocation' to teach and 'the year of routine pedagogy [at Valence] ended up destroying it' (21). Teaching practice was scarcely more inspiring since the teachers could not explain their methods, understand the beginner's efforts or correct mistakes. Looking back on the experience, Saunier declared: 'I had the impression of wasting my time. Precious time, since the war was finally over and I was twenty years of age' (21). A substitute was found for her teaching commitment and in the autumn of 1945 she moved to Paris in order to enter the University of Paris (Sorbonne). Having studied neither Latin nor Greek, she was not qualified to study most subjects in the arts and humanitics and dared not consider medicine or fine art; hence she had to choose between a degree in science or in geography. She opted for the latter and was taught at the Institut de Géographie by André Cholley, Georges Chabot, Charles Robequain, Jean Dresch and Aimé Perpillou (see *Geographers* 31 (Cholley, Dresch and Perpillou) and 32 (Robequain)). She also took a course in Russian at the Institut national de langues et civilisations orientales.

In October 1947 Alice Saunier married Elie-Jacques Picard, who was born on 5 December 1920 at Chagny (Saône-et-Loire *département*) and was educated in Lyon. He started clandestine activity for the Resistance as early as October 1940, distributing tracts and copies of the *Franc-Tireur* news sheet in the city and, eventually, as far afield as Clermont-Ferrand. In 1943 he joined the 'Kasanga' intelligence network and undertook many missions to report on German installations in Provence and along the Mediterranean coast, including the submarine base at Le Mourillon near Toulon. In June 1944 he was arrested by the Gestapo in Lyon and was interrogated by Klaus Barbie (the SS and Gestapo man known as the 'Butcher of Lyon' for his extremely brutal treatment of prisoners), before being imprisoned at Fort Montluc. From there he was deported to the Natzweiler-Struthof concentration camp in Alsace, and was then sent to Dachau, where his health deteriorated as a result of typhus and tuberculosis, as well as the harsh treatment. Jacques Picard was liberated by American troops on 29 April 1945 and returned to France a fortnight later. He set about preparing for an examination in mathematics to complement an earlier paper taken in the summer of 1942. He passed the new assessment in October 1945 but was in such poor health that he could not start university study at the same time as Alice. He spent much of the next three years in hospitals and sanatoria in Paris and in the mountains of Savoy. To help fund their studies, Alice worked as a personal assistant to a director of Electricité de France, the state-owned energy organization, during 1947 (Droit 1982; Bastié 2003, 101). This was not an entirely harmful experience, since she learned about office efficiency and how to cope with matters in a business-like fashion. She completed her first degree in 1948 and the couple moved to Innsbruck in Austria, where she held a lowly job in a *lycée* for children of the members of French forces of occupation in western Austria. This visit had important consequences for her future career and enabled her to start research on a *diplôme d'études supérieures* (equivalent to a master's degree by research). Unfortunately, her husband's health did not improve and in July 1950 the couple had to return to Paris where Jacques was readmitted to hospital. A few months later, Alice was also hospitalized because of tuberculosis and pleurisy. During a period of convalescence, Jacques studied part-time at the Sorbonne. Not until July 1951 were both Alice and Jacques discharged from hospital. Thereafter Jacques studied full-time for a degree in mathematics, which he obtained in 1952. At the same time, Alice completed her *diplôme d'études supérieures* with advice from Aimé Perpillou. It was devoted to the Brenner railway between Innsbruck and Bolzano.

Jacques commenced postgraduate studies in 1953, and Alice took a succession of teaching jobs, including being responsible for discipline at the Lycée Paul Bert and instructing teenagers in the rudiments of French language, history and geography at the Centre d'apprentissage at Savigny-sur-Orge in the southern suburbs of Paris. Working in ramshackle premises, Alice 'discovered urban adolescents whose harsh life and behaviour was totally different from that in the most disadvantaged parts of our countryside' (Saunier-Seïté 1982, 26). At this time, she joined the Confédération Générale du Travail (a communist trade union) through fraternity with fellow instructors. Because of her personal drive and manifest organizational skills, she 'rapidly acquired regional responsibilities in the union' but this commitment to the political left proved to be short-lived (26). As well as teaching and studying, the couple had two young sons, Philippe and Jacques, to look after. To their great sorrow,

Jacques was severely handicapped and, as was not uncommon at the time, Alice never talked about him in public (pers. comm. J.-R. Pitte, 30 March 2017). By the mid-1950s, Jacques senior had become an assistant lecturer in mathematics at the Faculty of Science of the Sorbonne. Alice grew increasingly frustrated by her teaching jobs and by 'the whole education policy in France that seemed to be demagogic improvisation, devoid of perspective and subject to the growing weight of trade unions and Marxism' (Saunier-Seïté, 1982, 26). The minimum school-leaving age was raised from fourteen to sixteen years at the same time as the largest cohort of the post-war 'baby boom' entered secondary school but, she believed, 'nothing was prepared to accommodate this: no statistical projections, no teaching programmes that could be adapted to the needs of pupils of very varied origins and levels, no plans for installation, or budgets for construction and equipment of additional premises' (26). The result seemed to be just 'incoherent and fatal improvisation', hence many pupils no longer knew how 'to read, write or count properly' or to have any sense of geography and history (Mazeaud 2007). As the months passed, she decided 'to quit the unsatisfactory school environment, with its bitterness, selfish security and incomprehension of the hard realities of life, as well as the nervous breakdowns and depression', in order to seek a new career (Mazeaud 2007).

As soon as her state of health allowed, Alice Picard began research for a state doctorate (*doctorat d'état*) that focused on the regional geography of part of the Tirol where she had spent some time teaching and recuperating after the war (Picard 1963). With strong support from Georges Chabot and from Raoul Blanchard (formerly at Grenoble University), she was appointed in 1958 to a research assistantship at the Centre National de la Recherche Scientifique (CNRS) that provided funding for five years (for Blanchard see *Geographers* 32). Then she was elected as secretary of the geography committee of the CNRS where she worked closely with its chairman, Blanchard, and had 'excellent relations' with his successor, Jean Dresch (Dresch 1989, 64). In this position she observed what she felt to be the best and worst characteristics of the French research community. The best involved 'the scientific passion and moral conscience of some of our senior professors', while the worst was 'the injustice . . . in awarding financial grants that not only reflected nepotism with regard to favourite disciples of certain all-powerful "bosses" (*patrons*), but increasingly mirrored the political colour of the applicants. At this time, one did well to be a Marxist', she remarked wryly (Saunier-Seïté 1982, 37). Recalling this period, geographer and pipe-smoker Pierre Monbeig (see *Geographers* 32) remembered Alice Picard as an 'excellent, charming and attractive colleague, who left a trace of perfume whenever she visited my office' (Picard 1991, 196) – a comment that tells us much about contemporary gender expectations and reveals her ability to win friends as well as to use the more confrontational tactics she later showed in her political career.

In 1963 Alice and Jacques were appointed to teaching positions at new institutions of higher education in the town of Brest in the far west of Brittany. Both colleges were offshoots of the University of Rennes. With his doctorate almost complete, Jacques taught mathematics at the Collège scientifique universitaire (CSU), which opened in 1959 (Fortuit 2007, 68). Having successfully defended her two doctoral theses in 1962 (published in 1963 and described in detail below), Alice was made senior lecturer (*maître de conférences*) in geography at the Collège littéraire universitaire (CLU), founded in 1960, where she operated under the tutelage of Professor André Meynier at Rennes (see *Geographers* 36). In the early days, conditions were hard at Brest since lecturers taught in prefabricated premises and lacked adequate library provision. As Alice recalled: 'We had to go to Rennes or Paris to borrow the books and maps we needed to prepare our classes' (Saunier-Seïté 1982, 38). Most lecturing colleagues were young and inexperienced. Support was given by local *lycée* teachers but they lacked any research ethos. Well-educated parents continued to send their children to university in Paris or Rennes, as they had always done, ignoring the new colleges at Brest that 'attracted students from less informed circles in the town and its surroundings' (40). Picard felt that the work and examination results of local students declined in quality year upon year in the mid-1960s (45). In January 1966 her colleagues requested the authorities in Rennes to appoint her director of the CLU. At the age of forty, she accepted this challenge and was soon elevated to a personal chair in geography at the University of Rennes, tenable at Brest. She proceeded enthusiastically with plans to build and equip new premises and to appoint additional academic and support staff to deal with increasing student numbers across the university (Lefeuvre 2011). She had a direct influence on young colleagues and oriented some to research in Iceland (Jean-Claude Bodéré) and the Arctic (Annick Moign) (1932–1987).

In the following year (1967) Picard was offered the post of *recteur*, with overall responsibility for all levels of state education in an *académie* (academic region). She declined this position, not wishing to abandon her fledgling CLU or interrupt her sons' education in Brest. A further invitation was to join the high-level Capelle Commission that had the objective of improving the standard of education in French universities. This time she accepted. Student unrest during May 1968 affected Brest as well as many other towns and Picard was confronted by strikes and requests from the union of students to stand down as head of the CLU. She stayed firm and remained in place throughout the summer, when Edgar Faure (1908–1988), the new minister of education, proposed a number of reforms regarding institutional governance, teaching staff, pedagogic methods, the creation of new universities and division of large old ones, notably the Sorbonne. These changes were duly incorporated into the orientation law, sometimes known as the 'Faure Law', of 12 November 1968 (Fridenson 2010). The CLU at Brest became part of the new University of Western Brittany, and Alice Picard was made dean of the faculty of arts, thereby retaining her position as director. She continued to press for additional buildings, equipment and staff. It was at this time that she met a general inspector of administration in national education by the name of Jérôme Seïté (d. 1972). He had worked closely with Faure in devising the reforms and became Alice Picard's second husband in 1968, following her divorce from Jacques Picard, who became a professor at the Université de Paris VI. Jacques Picard died in 2007.

In her role as director of the CLU and subsequently dean of faculty, Alice Saunier-Picard, Alice Saunier and Alice Saunier-Seïté as she finally styled herself, was pleased with her achievements at Brest, being especially proud of the research centre for Celtic and Breton studies, the laboratory for coastal geomorphology, the teacher training centre and the technical institute (Institut Universitaire de Technologie, or IUT) at Quimper. However, being in such a remote outpost of higher education continued to frustrate her (Saunier-Seïté 1982, 78). She sought advice from Georges Chabot, her doctoral supervisor, who informed her that an IUT was to be built at Sceaux as part of the new scientific university (Université de Paris XI) being established in the southern suburbs of Paris (83). In December 1969, Saunier-Seïté 'tore herself away, with considerable sorrow', but with pride in the 'open and collegial methods of working' that she had striven to implement at Brest during the previous six years (85).

Saunier-Seïté's new challenge at Sceaux was to have the college equipped, staffed and ready to receive students in the autumn of 1970, at 'one of the most beautiful sites in France' for a new institution (Mazeaud 2007). As director of the new IUT, Saunier-Seïté held a chair first at the Université de Paris V, and then at the new Paris XI where she served as a vice-president in the academic year 1970–1971. She was more than satisfied with the innovative teaching that she had managed to put in place at Sceaux but was infuriated by the disciplinary fragmentation of Paris XI, its seemingly interminable council meetings, and the many 'aberrant' decisions that led to 'grotesque errors' (Mazeaud 2007). In September 1972 her second husband died after less than four years of marriage. At the same time, her elder son started his training as a medical student, hence she found herself 'free, alone and ready for another absorbing and difficult task' (Saunier-Seïté 1982, 101; Choffel 1979). As her friend Pierre Mazeaud related: 'She immersed herself in work' (Mazeaud 2007).

Having declined two offers of a post as *recteur* (rector) Saunier-Seïté accepted the third invitation and set about managing state education in the *académie* of Reims, formed of five *départements* that had belonged to three pre-existing *académies* (Paris, Dijon and Lille). Examination results were below average in the region's schools, with particular challenges being posed by the needs of immigrant children and those who lived on canal barges. After leaving Sceaux in July 1973, Saunier-Seïté made her first visit to Reims in the following October and met five local inspectors of education who showed little prospect of working harmoniously. She discovered that the new University of Reims was dispersed in sites around the edge of the city, and was displaying a predictable range of early difficulties. Problems among its students, academics and support staff required firm but tactful responses from the rector. Her successes at Reims included establishing a boarding school for barge children, introducing mobile classrooms that brought scarce equipment to remote village schools, and extending technical education. The death of her father in 1975, less than three years after that of her husband, had an unsettling effect on Saunier-Seïté. Another career opportunity arose in the form of the rectorship of Corsica, where there was also a new university. She declined this offer, but remained emotionally attached to the island where her second husband

Jérôme had been mayor of Manso and she proceeded to serve as a municipal councillor from 1977 to 1983 (120).

On 12 January 1976 right-wing Prime Minister Jacques Chirac offered Alice Saunier-Seïté the position of secretary of state for universities, in succession to Jean-Pierre Soisson (Musselin 2004, 69). This move was made in full recognition of her tough stance as a university administrator. It was a clear acknowledgement of her personal capabilities and also responded to the campaign pledge of President Valéry Giscard d'Estaing to bring more women to high positions of government. Giscard had become acquainted with Saunier-Seïté during his visits to Brest and Reims. The two became 'very close', the president noting the impression that she made on entering the Council of Ministers in a miniskirt – an observation that reveals much about gender expectations among male politicians and about Saunier-Seïté's uncompromising determination (pers. comm. J.-R. Pitte, 31 March 2017). She accepted the challenge of managing 'the immense fleet' of higher education, comprising 750,000 students, 117,000 academic and support staff, 12 million square metres of buildings and one-sixth of the nation's civil budget. Despite some improvements since the student uprisings of 1968, she believed that French universities had subsequently 'gone through mutinies and storms that caused shipwrecks' (Saunier-Seïté 1982, 130). Chirac remarked: 'Do not thank me [for this position]. It is not a present,' and he confided later, 'I do not want you not to succeed where all of your predecessors have failed' (cited in Mazeaud 2007). Her main objective was to introduce measures to offset the disadvantages of reforms introduced hastily in the wake of the events of 1968. Saunier-Seïté's wide experience of educational administration convinced her that French universities had become highly politicized, 'good relations between teachers and students were destroyed, standards lowered, and effort negated; pursuit of equality was deadly for scholarship' (Mazeaud 2007). As Habiba Cohen observed: 'Her style was not one of accommodation. … She enjoyed jumping into the fray and polarizing a situation. When speaking in the National Assembly, she received loud ovations from the conservative benches' (Cohen 1978, 81). Few politicians escaped her criticisms and she did not spare female colleagues (pers. comm. J.-R. Pitte, 31 March 2017).

The new minister wasted no time and on 16 February she announced teaching reforms throughout state universities. These had been drafted by her predecessor but very much echoed her own mission to raise standards. Henceforth, universities were allowed to introduce innovative programmes of study, but the state retained responsibility for awarding – and withholding – degrees (Mohraz 2016, 120). Her announcement was greeted by protests from many students, academics and administrators who accused her of sacrificing educational solidarity for the kind of competition espoused by big business. Saunier-Seïté declared that 'Marxist unions of teachers and students made this text [of 16 January] their hobby-horse by misrepresenting its objectives' (Mazeaud 2007). A wave of agitation rocked universities until late April, but Saunier-Seïté was unwavering in her resolve. She declared that successive strikes damaged teaching and undermined the reputation of French higher education and the degrees it awarded. She agreed to halt the reforms if the president ordered her do so, but warned that dire consequences for the reputation of French higher education would follow. He gave no such instruction and the strikes faded away. Saunier-Seïté's firm stand paid off. Thereafter, she continued to press for improvements in teaching, learning and research, and sought to enhance the image of French universities at home and abroad.

In March 1977 Saunier-Seïté was made Minister for Universities in the government of Raymond Barre. She introduced further reforms involving technical education, strengthened the relationship between teaching and research, and fostered links between state universities and various branches of industry and commerce. Administrative and career structures in the CNRS were reorganized, provoking much opposition among bureaucrats and researchers. She favoured the modernization of existing university premises, rather than constructing new buildings. Revisions to the recruitment and working conditions of university instructors led to widespread controversy and strikes during 1978. Saunier-Seïté also had to manage the redistribution of teaching staff among universities in Paris, and to resolve the future of the Centre universitaire experimental de Vincennes that had been constructed in 1968 in woodland to the east of the capital owned by the City of Paris. Using prefabricated components, its buildings were erected within three months and received 8,000 students in the 1968–1969 academic session, already exceeding the maximum number they had been designed to accommodate. As well as providing much-needed extra places, the Centre universitaire was conceived as a locus of innovation in teaching and learning that was open to applicants without conventional qualifications (normally the *baccalauréat* certificate) as well as to those in work who wished to improve their education. In addition to covering many aspects of the arts and humanities, it introduced disciplines that were not usually taught in state universities, such as sociology, psychology, town planning, media studies, and creative and performing arts. It favoured less formal modes of learning than the traditional professorial lecture, and its students were encouraged to assemble their own programmes of study for assessment by other methods than the timed examination paper (Cohen 1978, 126–9).

Despite many real achievements, including provision of remedial classes and evening sessions for students in work, Vincennes seemed 'to have a disproportionate number of students with little initiative or sense of discipline' (Saunier-Seïté 1982, 128). It soon acquired a reputation as a place of confrontation dominated by various shades of the political left. This was only part of the picture, since there were also many 'ordinary' students, but it was the one that was carried repeatedly in the media (Soulié 2012; Idier 2012). In 1971 the centre acquired the title of Université de Paris VIII and the state endorsed the qualifications it awarded. A year later a convention specified that the site should be cleared of all buildings and returned to the City of Paris by 1978 at the latest (Saunier-Seïté 1982, 145). Jacques Chirac, as right-wing mayor of Paris (Rassemblement pour la République), was eager to have this controversial university re-established beyond his territory. Conditions became more chaotic as student numbers continued to grow, reaching 32,000 by 1976, of whom forty per cent lacked a *baccalauréat* qualification and almost seventy per cent were in full- or part-time work (Cohen 1978, 127). Half of both the undergraduate and postgraduate cohorts were made up of foreigners, many of whom had poor command of French. The seriously overcrowded institution simply lacked the necessary financial and teaching resources to cope. (Interestingly, Jérôme Seïté had been charged with investigating the university's monetary problems in the early 1970s.)

Reports of drug dealing on campus and student deaths due to overdoses added to the unfavourable publicity concerning Vincennes, to the consternation of Saunier-Seïté and the university authorities. In the summer of 1976 Saunier-Seïté initiated discussions on relocation and identified space in the new town of Marne-la-Vallée. This was rejected by the Council of Paris VIII and its president Pierre Merlin, a planner–geographer who saw the minister as 'our determined adversary' (Merlin 1980, 13). Various alternative locations were proposed, with Saunier-Seïté favouring a small (2 hectare) site at Saint-Denis in the northern suburbs, 15 kilometres from Vincennes and beyond the City of Paris. True to nature, 'She was directive and liked a fight' (Mazeaud 2007). Despite widespread protests, bulldozers razed the Vincennes campus during the summer of 1980 under surveillance from armed gendarmes. As Harald Wertz, lecturer in informatics, observed: 'Nothing remained less than a month after we left. Everything was demolished' (cited in Loubière 2001). Vincennes became 'a sort of martyr for the Left in French universities' (Le Cozanet and Dormoy-Rajramanan 2016, 283). *Le Canard enchaîné*, a satirical magazine, declared: 'Alice has lost her faculties' (cited in Wikipedia entry, Alice Saunier-Seïté). In the autumn, academics and students with appropriate qualifications were accommodated at Saint-Denis. With due irony, Saunier-Seïté remarked that this site was 'remarkably well located, being flanked by Lenin avenue, Stalingrad avenue, and Liberty street. This address seemed to be predestined … in a *département* [Seine-Saint-Denis] and in a town in the hands of the Communist Party, and whose politicians and press never cease to praise the exceptional merits of Vincennes' (Saunier-Seïté 1982, 145).

On 3 March 1981 Saunier-Seïté was given ministerial responsibility for women and families in addition to higher education. However, her role in government came to an end two months later following the election of socialist François Mitterrand as president. In the autumn of that year, it was proposed that she teach at the Conservatoire National des Arts et Métiers (national engineering school; CNAM) in central Paris. This suggestion caused initial controversy since her reputation as an uncompromising negotiator preceded her. In fact, academics at the CNAM were won over by her personal qualities and elected her professor of geography and planning. The clarity of her lectures and her unflagging organizational drive were much appreciated by her students and colleagues (Bastié 2003, 103). She remained at the CNAM until she retired in 1994, having chaired the Department of planning and spatial organization from 1986 to 1990. She once declared that these had been 'the best years' of her whole career (Mazeaud 2007).

Following her ministerial career, Saunier-Seïté served on the council of the City of Paris from 1982 to 2002, representing the 10th *arrondissement* (ward) and subsequently the 6th *arrondissement* in the latter of which she lived. She concerned herself initially with traffic, police and matters of security, and later chaired the commission for town planning (Rogister 2003). She was associate secretary general of the Parti Républicain from 1978 to 1982 and was active in the Mouvement national des élus locaux (local councillors), which she founded in 1978 and chaired during the 1990s. She convened important colloquia on the future of Europe and of the state in France, and brought her expertise to the Centre d'études et recherches sur Paris et l'Ile-de-France (regional planning), the Commission du Vieux-Paris (urban conservation), and the institution managing reservoirs and barrages in the upper Seine valley (Saunier-Seïté 1984, 1985). She belonged to the Société d'Histoire à Paris, the Société de Géographie de Paris, and the Club des Explorateurs. Her overseas interests were recognized by associate membership of the Icelandic Society for Sciences and the Institute of Egypt. In retirement, she was honoured by election to the Académie des sciences morales et Politiques of the Institut de France in 1995. She also found time to write historical studies of the Cardinal de Tournon (founder of the *lycée* where she studied), of the Courtenay family (from Burgundy), and of Count Boissy d'Anglas (who, like her, came from the Ardèche), as well as a book about recent politics (Saunier-Seïté 1980, 1997a, 1998a, 2001b). Alice Saunier-Seïté became 'rather whimsical' (*fantasque*) in her final years (pers. comm. J.-R. Pitte, 31 March 2017). She died of cancer in Paris on 4 August 2003, and was cremated four days later at a simple ceremony without eulogies at the cemetery of Père-Lachaise. In January 2017, the city council named the tiny Visconti-Saunier-Seïté Garden in the 6th *arrondissement* of Paris in her memory.

Geographical contributions by Alice Saunier-Seïté

Human adaptability to harsh environments formed a common thread in most of the publications of Saunier-Seïté, from her doctorate on part of the Austrian Tirol to an array of essays on Iceland, Scandinavia and northern Canada written throughout her career (Anon. 1997). Because of her administrative, ministerial and other political responsibilities, as well as the constraints of war, of bringing up children and dealing with poor health, she was not a particularly prolific author at certain periods of her career. She left the writing of textbooks for students to others and drafted only one geographical piece for a general readership – although, of course, her historical volumes were aimed at a broad market (Saunier-Seïté 1974a). Many of her articles on Iceland, Scandinavia and Canada appeared in *Norois*, the journal of geography departments in western France. Surprisingly, she did not publish in the *Annales de Géographie*, preferring to place early articles with *L'Information Géographique*, which catered mainly for schoolteachers, and later ones with *Acta Geographica*, the bulletin of the venerable Société de Géographie de Paris that published a collection of essays in her honour, following her retirement from the CNAM and her election to the Institut de France (Pitte 1997). Entitled *Apologie pour la géographie* (*The case for geography*), it responded to her wish to honour geography rather than her own achievements, and echoed the title of the famous book by historian Marc Bloch (1886–1944), *Apologie pour l'histoire, ou Le métier d'historien* (1949) (pers. comm. J.-R. Pitte, 31 March 2017). Its cover bears a photograph of the woods and fields of Saint-Jean-le-Centenier taken years previously by Jean-Robert Pitte.

Alice Picard wrote a cluster of articles on the agriculture, industry, tourism and transport facilities of the Austrian Tirol prior to her defence on 29 June 1963 of her major thesis about the northern valleys of the Oetztal upland (Picard 1958a, b, 1961a, 1962a, b, 1963, 1964a). The thesis comprised a 475-page volume of text with thirty photographs, and a minor volume containing almost eighty maps. Prepared under the guidance of Georges Chabot, it was a holistic regional study based on 'thirty visits spread over a decade', involving numerous surveys in the field and intensive work in the archives and libraries of Innsbruck (Picard 1964b, 184). The work 'astonished the Austrians', since it covered only the northern valleys of the upland and not the physical region as a whole that was the usual framework for geographical investigation (Picard 1963, 7). In fact, many previous enquiries by Austrian scholars focused sharply on specific localities. Divergent watercourses, high mountain peaks and a harsh Alpine climate, composed of a host of microclimates that reflected elevation, orientation and exposure, made this a region of difficulty for human occupation. Nonetheless, it remained settled and economically vibrant, unlike many parts of the French Alps, studied by Raoul Blanchard and his disciples at Grenoble, that were characterized by 'settlement desertion and ageing of the remaining population' (7). In the words of Picard, 'the natural environment of the Oetztal massif is both sumptuous and miserly. Only tourism can draw genuine prosperity from its landscapes . . . Its physical imperatives . . . tyrannise mankind without respite' (13). In five chapters she depicted the 'severe, even repulsive conditions' of the upland, whose farmers not only occupied its 'rare basins and flat land' but also 'infiltrated the harsh and uncertain margins of the high valleys' (89).

In the remaining four-fifths of her thesis, Picard presented carefully classified aspects of human activity in the northern Oetztal. She showed how medieval clearance had pushed up from the valleys on to higher surfaces, establishing a complex array of seasonal pastures on former stretches of woodland. In her own words: 'Cropping and livestock rearing were never sufficient to feed the abundant population that always looked to complementary resources, including craft activities and temporary outmigration' during the winter months to parts of southern Germany, with migrants returning to their holdings during the more clement part of the year (208). In the 1930s many Germans came to the Tirol for winter sports and during the German occupation of Austria from 1938 to 1945, the valleys received an important flow of investment in support of agriculture and traditional metal working, especially at Fulpes, and the provision of accommodation for visitors. Occupation of the area by French troops immediately after the Second World War led to a temporary decline but activities recovered after 1950, thanks to military indemnities and loans that led to 'a spectacular regeneration of tourism equipment and the transformation of villages in the Tirol, especially in the Oetztal upland' (385). Improvements to roads and transport, and provision of electricity certainly played their part, but the most important factor was the remarkable initiative taken by local people. As a result, the presence of tourists in both summer and winter became 'the most striking feature in the geography of the Oetztal massif' (343). Picard concluded: 'It is tourism that has transformed the Oetztal: the incredible flourishing of new and renovated buildings, [and] the installation of new roads allowing an inflow of vehicles and visitors are certainly striking. This reveals an opulence, even prodigality, which contrasts strongly with the mediocre resources of the land' (416).

Picard's major thesis was reviewed by two senior geographers who knew her well through her work at the CNRS and later at Brest. Raoul Blanchard stated that the thesis was 'of powerful interest, with several new features' (Blanchard 1964, 343). He was especially impressed by the author's very detailed analysis of the seasonal use of high pastures and her exploration of the role of local investors in stimulating both summer and winter tourism. In fact, Picard had commented on both these topics at the 19th International Geographical Congress in Sweden in 1960 (Chabot 1961, 466). Blanchard concluded that 'in these respects, the book of Madame Picard is revolutionary, and is recommended strongly to all concerned with problems of Alpine geography' (Blanchard 1964, 345). André Meynier felt that the thesis was 'classically inspired by the spirit and methods of regional geography', and was especially successful in drawing together information from various disciplines 'in whose harmonisation the originality of the geographer resides' (Meynier 1964a, 76, 78). Writing from Grenoble, Germaine Veyret-Verner offered a more nuanced appraisal (On Veyret-Verner see *Geographers* 37). She regretted that Picard had not visited the Institut de Géographie Alpine and that she appeared to be unaware of many of the books and articles that its members had written about mountain geography. She argued that such exposure 'might have allowed her to see Austrian work in a new light' (Veyret-Verner 1964, 373). Similarly, Picard would have benefitted from contact with Grenoblois geologists. While her bibliography was impressive for the volume and range of Austrian sources, there were 'numerous omissions' of relevant French work (373). The reviewer felt that the northern Oetztal, with only 25,000 inhabitants, was an exceptionally small area for a doctoral thesis. Despite its shortcomings, Picard's work displayed 'sincere enthusiasm and has the great merit of offering an excellent example of changing life in a mountain area. It provides very interesting points of comparison with our own knowledge of the [western] Alps' (374). Veyret-Verner concluded by thanking the author for a thesis that 'will be welcomed by Alpine geographers' (372).

As was the requirement of the time, Picard also wrote a minor thesis covering a different part of geography, in order to demonstrate her competence in – and ability to teach – the whole of the subject. Her choice was an analysis of the occurrence of the *Föhn* wind in the Innsbruck area during a five-year period from January 1957 to December 1961. Compiled with advice from Pierre Birot (1909–1984; see *Geographers* 34) and Pierre Pédélaborde (1911–1992) at the Sorbonne and from Austrian meteorologists, it added to earlier work by local scientists, and in due course was extended by later findings from her own students (Picard 1967a; Saunier-Picard 1969a). Building on her general discussion of the warming *Föhn,* Picard presented a longer and fully illustrated account in a series published by the CNRS (Picard 1958c, 1965a). It was praised for its original exploration of the atmospheric conditions that triggered the occurrence of the *Föhn* wind, and its discussion of 'the relative importance of different levels of the troposphere in the genesis of climatic features' close to the surface of the Earth (Carrière 1965, 565; Meynier 1964b). In 1963, the Société de Géographie de Paris awarded its Jean-Baptiste Morot Prize (silver medal) to Picard for her doctoral work.

Even before her work on the Tirol was complete, Picard visited Iceland, where she undertook her first fieldwork in the summer of 1958, and subsequently other harsh environments in Scandinavia and North America (Picard 1964c, d, e, f, g). She attended colloquia on polar research and established particularly close links with scholars in Iceland, becoming a friend of President Vigdis Finnbogadóttir (b.1930), the world's first elected female president, in post from 1980 to 1996. A suite of articles from Picard presented the changing physical and human geography of that remarkable island, with texts on volcanic eruptions, ecology and territorial waters being complemented by discussions of agriculture and rural life, economic planning, population change and other themes (Picard 1959a, b, c, d, 1961b, 1967b, c; Saunier-Picard 1968, 1973, 1974b). Many of these essays appeared in *Norois*, which also published her accounts of Arctic and Subarctic environmental research, of polar investigations by German scientists, and of settlement, economic development and indigenous peoples in northern Canada (Picard 1965b, 1966, 1967d; Picard and Péguy 1965; Saunier-Picard 1969b, 1970a; Saunier 1970b, c, 1971, 1973; Saunier-Seïté 1974b, 1975, 1987a). It is perhaps regrettable that she did not place more of her 'northern' writings in more prominent geographical or interdisciplinary journals. Her commitment to the Société de Géographie de Paris was demonstrated by papers in its bulletin on topics as diverse as historical geography, past meteorological conditions, an Egyptian oasis, disease in topical environments, the conservation of the Bièvre valley in Paris, and, of course, the geography of Iceland (Saunier-Seïté 1970d, 1987b, 1995, 1996, 1997b, 1998b, 2001c). Her last publication was an affectionate essay in memory of fellow geographer René Haby (1919–2003) who also came from working-class origins and rose to be minister of education. The two friends shared the 'ideal and noble task' of improving the learning opportunities of young people in France (Saunier-Seïté 2003, 94).

Conclusion

The life of Alice Saunier-Seïté was remarkable in many ways. Coming from a humble family background, she coped with her own ill health and that of her first husband, and the challenges of caring for a handicapped child. She became the eighth woman in France to obtain a state doctorate in geography, behind such pioneers as Thérèse Sclafert (awarded 1926), and Jacqueline Beaujeu-Garnier (1947; see *Geographers* 28) and Germaine Veyret-Verner (1948) (Joly 2013; Ginsburger 2017). Her early involvement with the CNRS revealed unwavering commitment to efficiency and improvement, a quality that she put to good advantage throughout her career. As Michel Droit remarked, contrasting stereotypical views: 'This pretty, elegant, witty and attractive woman' was also 'a fighter' (Droit 1982, 7–8). She was the first woman to become a faculty dean in a French university, and the first female rector of an *académie*. She was not the first woman to hold ministerial office (being third) but was the first woman to hold responsibility for higher education. Having left government, she was the first woman to be elected a professor at the CNAM, and the first member of the Paris city Council to be elected to the Institut de France. In addition to all these responsibilities, she spent periods of her life as a school teacher, university lecturer and professor. She wrote an array of good but strongly empirical papers that arguably reflect the nature of her early training for the teaching profession. Their value is to be found in their factual content rather than in any form of methodological innovation.

Alice Saunier-Seïté's was a 'complicated personal and professional life' (pers. comm. Ian Thompson, 9 April 2017) and one in which she was often judged by different standards from her male peers. She is remembered more for her role as a minister than as a geographer and assessments of her time in office are necessarily coloured by commentators' gender and own political orientation. There is widespread agreement that, from the moment of her appointment as secretary of state for universities, she displayed unfailing dedication to her responsibilities. In a judgement couched in gendered terms, Michel Droit declared: 'She became the "strong man" of the government. Some admired her, even loved her. And what is perhaps an even better sign, some feared her, even detested her ... For five years she was on the front line, sometimes receiving as many blows as she gave out, but never losing sight of her objective: to contribute to giving France a university system . . . for the twenty-first century' (Droit 1982, 9). As she sought to mitigate the disadvantages of the hastily introduced 'Faure Law', she 'brought considerable strengths of character, administrative skills, diplomacy, charm and, above all, vision to the task. She wanted French universities to be poles of excellence ... She made their libraries more research-oriented and created new faculties. Reforms of the CNRS and French cultural institutes abroad were also put in hand' (Rogister 2003). Of course, all this activity and her political orientation earned her enemies as well as friends in higher education. Mention of her name is never greeted with indifference by middle-aged or older academics in France.

Following the announcement of her death, Jean-Pierre Raffarin, France's sometime centre-right prime minister, hailed Alice Saunier-Seïté as 'one of the major actors who opened French universities to modernity'; Alain Juppé, also a centre-right prime minister, spoke of her 'courage and determination in defence of her convictions'; and Jacques Chirac praised the 'tenacity and conviction' of this 'woman of great courage in her personal life and her professional career'. Fencing with foil and sabre served as an outlet for her aggression (Rogister 2003). She also qualified to pilot a light aircraft, but Chirac banned that hobby when she held ministerial office. An old friend, geographer Jean Bastié (1919–2018), recalled that she was

> a great believer in free and frank expression, never mincing her words. She criticised her friends as often as her adversaries, but she knew how to recognise the merits of both camps. She was generous and faithful in friendship, and full of good humour. She retained her enthusiasm and ardour to the very end. A journalist once described her as 'Chirac in a petticoat'. Nothing gave her more pleasure, even if she was well aware of the President's weaknesses (Bastié 2003, 103).

As a politician, Alice Saunier-Seïté contributed substantially to the modernization of universities in France, even if her methods were sometimes brutal and her approach confrontational. Her political style made her plenty of enemies – as well as friends. Likewise, her legacy as a geographer is mixed. The impact of her research was limited, even though her publications are fairly numerous for someone who was not primarily a researcher. Her teaching is perhaps best remembered through her later years at the CNAM. As an academic administrator and leader, however, her legacy is substantial in the universities of Brest, Reims and Paris XI. Her experience within geography and as a university administrator evidently shaped educational reforms on the wider stage of French universities. Without this experience within academic geography she would neither have been invited to hold high office nor have used her time in office as she did.

Acknowledgements

My thanks are extended to Ian Thompson and the late François Gay for their encouragement and I am most grateful to Jean-Robert Pitte who generously shared with me his recollections of Alice Saunier-Seïté.

Note

The current titles of the three universities in Saunier-Seïté's career are Université de Bretagne Occidentale (Brest), Université de Reims-Champagne-Ardenne, and Université Paris-Sud (Paris XI).

Sources

Saunier-Seïté discussed her life and work up to the end of her ministerial responsibilities in *En premiere ligne* (1982). A valuable review of her activities was presented to the Institut de France on 25 January 2007 by Pierre Mazeaud (2007) httpsi//www.asmp.fr/travaux/notices/mazeaud_saunier.htm. Information on Saunier-Seïté as rector of Reims *académie* is provided at http://www.cndp.fr/crdp-reims/ressouces/dossiers/bicentenaire-rectorat/pannneau17.htm Video clips of Saunier-Seïté may be found at www.ina.fr/

video/10007000 and www.ina.fr/video/CAB7600510801 A biography of Elie-Jacques Picard is to be found at http://www.fondationresistance.org/pages/rech-doc/?p=portaits&ildPortrait=17# Saunier-Seïté's personal file is lodged at the Archives Nationales 100/ AJ.

References to the academic and political milieu of Alice Saunier-Seïté

Anon. (1997), 'Bibliographie des travaux de Alice Saunier-Seïté', in J.-R. Pitte (ed.), *Apologie pour la géographie*. Paris: Société de Géographie, 7–10.

Bastié, J. (2003), 'Alice Saunier-Seïté, 1925–2003', *La Géographie* 1511, 100–3.

Blanchard, R. (1964), 'Une étude de montagne en Tirol', *Annales de Géographie* 73, 343–5.

Bloch, M. (1949), *Apologie pour l'histoire, ou Le métier de l'historien*. Paris: Armand Colin.

Carrière, P. (1965), 'Une importante contribution à l'étude du Suedfoehn d'Innsbruck', *Bulletin de la Société Languedocienne de Géographie* 36, 559–65.

Chabot, G. (1961), 'Géographie humaine', *Annales de Géographie* 70, 462–9.

Choffel, J. (1979), *Seule une femme: Alice Saunier-Seïté*. Paris: Flammarion.

Cohen, H. S. (1978), *Elusive Reform: The French Universities, 1968–1978*. Boulder: Westview.

Dresch, J. (1989), 'La géographie', *Cahiers pour l'Histoire du CNRS* 2, 55–69.

Droit, M. (1982), 'Préface', in A. Saunier-Seïté (ed.), *En première ligne*. Paris: Plon, 9–12.

Fortuit, J.-C. (2007), 'Alice Saunier-Seïté et Charcot', *La Géographie* 1527 (hors-série numéro 1), 67–70.

Fridenson, P. (2010), 'La politique universitaire depuis 1968', *Le Mouvement Social* 233, 47–67.

Ginsburger, N. (2017), 'Portrait en groupe de femmes–géographes. La féminisation du champ disciplinaire au milieu du XXe siècle, entre effets de contexte et de structure (1938–1960)', *Annales de Géographie* 116, 107–33.

Idier, A. (2012), 'Un mythe à détruire', *Lectures*, http://lectures.revues.org/8256

Joly, G. (2013), *Dictionnaire biographique de géographes français du XXe siècle, aujourd'hui disparus*. Paris: Grafigéo.

Lefeuvre, S. (2011), 'La filière AES, un renouvellement de la professionnalisation et de la démocratisation dans l'Université française?', *Cahiers de la Recherche sur l'Education et les Savoirs* numéro hors-série 3, 111–28.

Loubière, P. (2001), 'Paris–VIII la fac pas comme les autres', *Le Nouvel Observateur*, 31 May.

Mazeaud, P. (2007), *Notice sur la vie et les travaux de Alice Saunier-Seïté*. Paris: Institut de France.

Merlin, P. (1980), *L'université assassinée: Vincennes, 1968–1980*. Paris: Ramsay.

Meynier, A. (1964a), 'Les vallées septentrionales du massif de l'Oetztal', *Norois* 41, 76–8.

Meynier, A. (1964b), 'Contribution à l'étude du Sudföhn d'Innsbruck', *Norois* 41, 78–9.

Mohraz, N. (2016), 'L'UNI (Union nationale interuniversitaire): l'opposition de droite la plus virulente à la Loi Faure, 1968-1984', in B. Poncet and D. Valence (eds), *La Loi Edgar Faure. Reformer l'université depuis 1968*. Rennes: Presses Universitaires de Rennes, 99–121.

Musselin, C. (2004), *The Long March of French Universities*. New York: Routledge.

Picard, J.-F. (1991), 'Pierre Monbeig et le CNRS', in M. Droulers and H. Théry (eds), *Pierre Monbeig, un géographe pionnier*. Paris: Editions de l'Institut des Hautes Etudes de l'Amérique Latine, 193–205.

Pitte, J. R., ed. (1997), *Apologie pour la géographie. Mélanges offerts à Alice Saunier-Seïté*. Paris: Société de Géographie.

Rogister, J. (2003), 'Alice Saunier-Seïté: A woman in Napoleon's footsteps', *The Guardian*, 13 August.

Soulié, C. (2012), *Un mythe à détruire ? Origines et destin du Centre universitaire expérimental de Vincennes*. Saint-Denis: Presses Universitaires de Vincennes.

Veyret-Verner, G. (1964), 'Les vallées septentrionales du massif de l'Oetztal', *Revue de Géographie Alpine* 52, 372–4.

Selected publications by Alice Saunier-Seïté (Those prior to 1968 are shown as by Alice Picard, and others between 1968 and 1974 as by Alice Saunier-Picard*, or simply by Alice Saunier#)

1958a 'L'industrie de fer dans le Stubaital', *Bulletin de l'Association de Géographes Français* 278–9, 51–62.

1958b 'Quelques remarques sur la culture du maïs dans le Tirol autrichien', *L'Information Géographique* 22, 72–4.

1958c 'Connaissance du Föhn', *L'Information Géographique* 22, 209–20.

1959a 'La biogéographie des eaux islandaises et ses incidences économiques et politiques', *L'Information Géographique* 23, 68–72.

1959b 'La limite des eaux territoriales islandaises', *Annales de Géographie* 68, 90–2.

1959c 'L'habitat rural en Islande', *Revue du Nord* 163, 21–31.

1959d 'L'élevage et la politique agricole de l'Islande', *L'Information Géographique* 23, 154–63.

1961a 'Les paysans et le tourisme dans le Tirol autrichien', *L'Information Géographique* 25, 101–3.

1961b 'Vestmannaeyjar', *L'Information Géographique* 25, 65–8.

1962a 'Le Brenner et son autoroute', *Revue de Géographie de Lyon* 37, 357–63.

1962b 'Rôle des paysans dans l'aménagement touristique des hautes vallées tiroliennes', in *Comment concevoir et animer une économie du Massif Central.* Clermont-Ferrand: Faculté des Lettres, 153–80.

1963 *Les vallées septentrionales du Massif de l'Oetztal*, 2 vols. Paris: SEDES.

1964a 'L'exemple des hautes vallées tiroliennes', *Bulletin du Comité National de Géographie* 3, 27–34.

1964b 'Les vallées septentrionales du massif de l'Oetztal', *L'Information Géographique* 28, 183–4.

1964c 'L'essor récent des transports islandais', *Norois* 42, 205–12.

1964d 'Les problèmes du Nordland', *Internord* 6, 32–41.

1964e 'Programme économique islandais', *Internord* 6, 98–105.

1964f 'L'économie suédoise au tournant de 1963–1964', *Internord* 7, 55–83.

1964g 'L'archipel des Féroé', *Internord* 7, 119–31.

1965a 'Contribution à l'étude du Sudföhn d'Innsbruck', *Centre National de la Recherche Scientifique: Mémoires et Documents Cartographiques et Géographiques* 10(1), 1–116.

1965b 'L'océan glacial arctique et les pays du Nord: la péninsule scandinave, l'Islande, le Groenland, le Canada, l'Alaska, l'Arctique soviétique', *Historiens et Géographes* 94, 244–59.

1965c With Péguy, C.-P., 'Chronique arctique', *Norois* 48, 501–26.

1966 'Ecologie de la zone subarctique', *Norois* 52, 635–54.

1967a 'Quelques observations sur les apparitions du sud föhn à Innsbruck', *Publications de l'Institut Suisse de Météorologie* 4, 253–4.

1967b 'Recherches sur la nouvelle ile volcanique de Surtsey, dans la Baie du Roi et en Finlande', *Norois* 56, 635–50.

1967c 'L'Islande rurale', *Norois* 53, 58–67.

1967d 'Les sixièmes journées polaires de la Société allemande de recherches polaires', *Norois* 56, 650–65.

1968 'Population et économie de l'Islande', *Norois* 60, 545–67.*

1969a 'Le föhn à Altdorf et à Innsbruck en 1963, 1964 et 1965', *La Météorologie*, 199–216.*

1969b 'Enquêtes sur les agglomérations du Nord canadien', *Norois* 64, 607–34.*

1969c 'Notes sur les climats polaires et subpolaires', *Norois* 61, 63–83.*

1970a 'Quelques aspects de la géographie humaine du Nord canadien: origine du peuplement, découverte et mise en valeur du Canada septentrional', *Norois* 68, 649–62.#

1970b 'Les Esquimaux du Canada', *Norois* 69, 143–56.#

1970c 'Les septièmes journées polaires de la Société allemande de recherches polaires', *Norois* 65, 99–109.*

1970d 'L'Islande d'hier et d'aujourd'hui', *Acta Geographica* 3, 34–47.

1971 'Nouvelles d'Islande ; huitième congrès de la Société allemande de recherches polaires ; mise en valeur des régions arctiques et subarctiques', *Norois* 72, 739–64.#

1972 'Biométéorologie humaine et pathologie géographique', in *La pensée géographique française contemporaine. Mélanges offerts au professeur A. Meynier*. Saint-Brieuc: Presses Universitaires de Bretagne, 381–93.

1973 'Heimaey, la nouvelle Pompéi', *Norois* 80, 717–26.#

1974a 'Spitsberg, Groenland, Iles Féroé, Islande, Scandinavie, Finlande', in P. Deffontaines, M. Jean-Brunhes Delamarre and A. Journaux (eds), *Géographie régionale: Encyclopédie de la Pléiade*, vol. 1. Paris: Gallimard, 585–710.

1974b 'La guerre de la morue', *Norois* 81, 146–55.

1975 'Climat et pathologie', *Evolution Médicale*, 287–91.

1980 *Le Comte Boissy d'Anglas: Conventionnel et pair de France*. Paris: France Univers.

1982 *En Première Ligne: De la communale aux universités*. Paris: Plon.

1984 ed. *Remettre l'état à sa place*. Paris: Plon.

1985 ed. *L'Europe à la carte*. Paris: Plon.

1987a 'Les recherches françaises en Islande', *Cahiers des Explorateurs* 26, 19–22.

1987b 'A propos de la tempête qui causa le naufrage du Pourquoi Pas?', *Acta Geographica* 69, 12–9.

1995 'De la géographie grecque à la cartographie occidentale du XVIIe siècle', *Acta Geographica* 101, 3–18.

1996 'Siouah, oasis égyptienne du désert lybien', *Acta Geographica* 108, 3–18.

1997b 'Islande, le feu sous la glace', *Acta Geographica* 109, 5–29.

1997a *Le Cardinal de Tournon, le Richelieu de François Ier*. Paris: Bartillat.

1998a *Les Courtenay: Destin d'une illustre famille bourguignonne*. Paris: France Empire.

1998b 'Pathologie géographique: sur quelques endémies intertropicales', *Acta Geographica* 115, 41–53.

2001a *Le Comte Boissy d'Anglas. Conventionnel et pair de France*, 2nd edn. Paris: France Univers.

2001b *Giscard à deux voix*. Paris: Perrin.

2001c 'Revoir la Bièvre à Paris', *La Géographie* 1502, 59–66.

2003 'René Haby, 1919–2003', *La Géographie* 1509, 91–4.

Chronology

1925	Born on 26 April at Saint-Jean-le-Centenier, in the Ardèche
1937–41	Educated at the Ecole primaire supérieure at Privas
1941	Passed entrance examination for teacher training at the Lycée de Tournon
1944–5	Studied at teacher training college at Valence
1945–8	Studied at the Sorbonne in Paris, graduating with degree in geography
1947	Married Elie-Jacques Picard
1948–9	Taught at French school near Innsbruck
1951	Left hospital
1952	Completed *diplôme d'études supérieures*
1952–8	Taught at various schools in and around Paris; worked as personal assistant
1958–63	Undertook research funded by the CNRS; worked as secretary to the chair of the geography committee of the CNRS
1962	Awarded *doctorat d'état*
1963	Publication of *Les vallées septentrionales du Massif de l'Oetztal;* received the Jean-Baptiste Morot Prize
1963–5	Made senior lecturer at the Collège littéraire universitaire in Brest; member of the Permanent Section of the CNRS
1966	Elevated to a personal chair in geography at Brest
1966–8	Served as director of the Collège littéraire universitaire
1967–9	Elevated to an established chair of geography at Brest
1968–9	Elected dean of the arts faculty of the University of Western Brittany
1968	Married Jérôme Seïté, having divorced Elie-Jacques Picard
1969–70	Made professor at the Université Paris V; founder of the IUT at Sceaux
1970–71	Served as vice-president of the Université Paris XI
1970–73	Professor at the Université Paris XI; director of the IUT at Sceaux
1973–76	Rector of the *académie* of Reims
1976–78	Served as Secrétaire d'Etat aux Universités
1976	Elected associate member of the Institut d'Egypte
1978–81	Served as Ministre des Universités
1978–82	Assistant general secretary of the Parti Républicain

1981	Appointed Minister for the Family and Women
1981–94	Professor at the Conservatoire National des Arts et Métiers
1982	Elected associate member of the Icelandic Academy of Sciences
1983–95	Served on the council of the City of Paris
1990–95	Made treasurer then president of the Institut Océanographique in Monaco
1990–98	Served as President of the Mouvement national des élus locaux
1992–99	President of the Mutuelle de la caisse de retraite and of the Institut de formation des élus locaux
1995	Elected member of the Académie des sciences morales et politiques, Institut de France
1995–98	President of the Centre d'études et de recherches sur Paris et l'Ile-de-France
2003	Died in Paris on 4 August; cremated at the cemetery of Père-Lachaise on 8 August

4

Helen Margaret Wallis, 1924–1995

Sarah Tyacke

Figure 4.1 Helen Wallis at her desk in the British Library (BL) Map Library with the catalogue of the Drake exhibtion 1977 before her (photo: priv. coll.).

Helen Margaret Wallis's death in 1995 at the relatively early age of 70 was a great loss to many academic worlds: the history of cartography, the history of travel and exploration, globe history and map librarianship, all of which may be said to fall within a broad definition of historical geography. Her loss was felt even more by the international community of scholars and professionals and by the many societies Wallis either helped found or ran in her long career from 1951 to her death. She was interdisciplinary before it became fashionable, and this account explores in more detail her life, work and scholarly and professional contributions.

Early life and education

Helen Wallis was born on 17 August 1924 at Dunkery, Park Road, Barnet, Hertfordshire, the daughter of Leonard Francis Wallis (1880–1965), headmaster, and his wife, Mary McCulloch Jones (1884–1957), schoolteacher (Campbell 2004). She had a twin brother Peter, a defence research scientist who is still alive. Her long-standing friends Catherine Carver, a fellow pupil at St Paul's School, London, and Brenda Hall, a fellow student at St Hugh's College, University of Oxford, recall that it was from her parents that she was imbued with integrity and high principles combined with a generous and tolerant disposition to all those around her (Tyacke and Campbell 1995, 14–15, 79–80). Indeed, she rarely made disparaging comments about her friends or colleagues, whatever she may have felt on occasions.

Helen Wallis attended the fee-paying preparatory school Colet's Girls' School, London, and then its associated secondary school, the highly academic St Paul's School for Girls, London, in 1934 on a junior foundation scholarship. She stayed at St Paul's until 1943 after which she spent the rest of the war teaching at Colet's Girls' School. Both geography and history held her interest at St Paul's, but geography won, she said, because the history teacher wanted 'clear exposition' rather than scholarship – something that Helen thought was writing 'like a journalist'. While at St Paul's, Helen won many school prizes, in particular the Castlecomer classics prize (1940) and the Alice Lupton Senior Piano prize (1942) (Tyacke and Campbell 1995, 13–14) She remained a lifelong friend to St Paul's. Through her long relationship with Davidson College, North Carolina, she was instrumental in enabling pupils from St Paul's to apply for the Dean Rusk scholarship there. This came about because of her great friendship with Professor W. P. (Bill) Cummings and his wife Betty when she was at Davidson in 1985 receiving the honorary degree of DLitt (British Library Helen Wallis archive (hereafter BL HWA): letter to HW from the High Mistress, Heather Brigstocke 6 Oct 1983). In her will she left one of her treasured copies of Jean Rotz *The Boke of Idrography 1542* to St Paul's (Bailes 2000, 134–5).

From St Paul's Wallis went to St Hugh's College, Oxford, on an exhibition (a minor scholarship), to read for a BA degree in geography at the Oxford School of Geography. She worked particularly under J. N. L. Baker who was a renowned historical geographer and who had published on the history of explorations. She won various awards while at Oxford: in particular, her essay 'An application of the principle of regionalism to the study of English administrative units: a region in Leicestershire' – a region chosen perhaps because her father's home town was Leicester – won the Royal Geographical Society (RGS)'s university essay prize (BL HWA, early works and letter 23 June 1948). The award of the RGS prize marked the start of a long and close association with that society. As an undergraduate Wallis went on field trips to the Pyrenees and to the Austrian Alps and Vienna and on each, as became characteristic, she took many photographs. Vienna was particularly attractive to her as a photographer, being at that time under the occupation of the four victorious allies (USA, UK, France, USSR) after the Second World War. A photograph of the changeover of French and Russian occupation troops is marked on the back: 'Back view of a Russian officer, whom I was warned not to photograph' (BL HWA, *c.*1950 album of photographs).

Wallis graduated BA degree in 1949 and election to the Mary Gray Allen Senior Scholarship at St Hugh's in 1949 (Bailes 2000, 134–5) enabled her to pursue research. With interests in both history and geography, on the advice of Baker, she embarked on an Oxford doctoral thesis under the supervision of R. A. Skelton, superintendent of the British Museum (BM) Map Room (1950–1967), an expert on early maps, travel and exploration and honorary secretary of the Hakluyt Society, which published journals of discovery and exploration (Interview with T. Campbell and S. Tyacke 1994, hereafter 'HW video 1994'). Wallis's thesis was on 'The exploration of the South Sea, 1519 to 1644' (1954a) and she examined not just the routes of the Pacific explorers, but also the climatic influences on those routes, and how far the explorers succeeded in achieving what they had set out to accomplish. The Pacific remained of critical interest to Wallis thereafter. Her thesis was examined by J. N. L. Baker and the maritime historian J. A. Williamson who was an expert in the voyages of the Cabots. According to her friend Brenda Hall, then in the University Registrar's office, it was thought at the time to be of very high quality, and Andrew David, the editor of *Cook's Charts and Views* shared that view some decades later (Tyacke and Campbell 1995, 14; St Hugh's 1997b, 28–30).

Wallis remained devoted to St Hugh's College, becoming President of the Association of Senior Members in 1982 and thus in office during the college's

Figure 4.2 A commemorative glass plaque for Helen Wallis, engraved by Geoffrey Thwaites, was placed on the west wall of the chapel of St Hugh's College on 27 June 1996, given by her friends (1996, pers. comm. and photo by A. D. Baynes-Cope).

centenary celebrations. While the two were at St Hugh's, Brenda Hall remembers Helen's being baptized into the Church of England and Wallis remained a devoted Anglican thereafter (St Hugh's 1997a, 15). She occasionally played the organ in the chapel at St Hugh's, continuing a love of music, which had been fostered at school when she was made Licentiate of the Royal Academy of Music for piano performance.

Early career and the BM Map Room 1951–1967

Wallis's father was a great influence on her life until his death in 1965. He was a lifelong member of the Fabian Society (of left-wing intellectuals) and Helen joined the young Fabians; she remembered her father talking about Bernard Shaw, Sidney and Beatrice Webb and H. G. Wells – all progressive figures in different ways (BL HWA 'Memorial service for Leonard Francis Wallis 21 Nov 1965' and associated papers). After her father's retirement in 1947 he drove her to the various aristocratic and country houses across the country where, in 1949, she began to compile the first entries of the catalogue of all the terrestrial and celestial globes in this country for the Commission de Bibliographie, Union Internationale d'Histoire des Sciences; under the initial supervision of

Edward Lynam (*d.*1950), then Superintendent of the Map Room but shortly to retire. Lynam wrote that he had 'engaged as my assistant Miss Helen Wallis … of St Hugh's College Oxford. Miss Wallis was engaged upon a thesis on the history of exploration … but agreed to work in her spare time. I drew up for Miss Wallis some simple rules for cataloguing globes. After Christmas 1948 she began work under my supervision on the globes of the British Museum. She has proved a most efficient and enthusiastic worker' (BL HWA 'Lynam to Chair of Commission' 1 Sept 1949). Thus began Wallis's long association with the BM Map Room and these early influences remained with her for the rest of her life. It seems that Baker had hoped Wallis might take up a junior tutorship at St Hugh's, which was to be established. When she wrote to Baker to ask for a reference he mentioned this possibility to her and wondered if she would really 'wish to spend the rest of her life sitting in the British Museum' (HW video 1994): but, as Wallis recorded in her interview with Tony Campbell and Sarah Tyacke in August 1994, she had worked in the Map Room and 'thought it a very good place to be'. She said that she had never regretted her choice, since in some ways, she thought, she would have felt more constrained in Oxford, although she remained devoted to St Hugh's. Certainly, had she gained the St Hugh's post she felt that she would never have fulfilled her own scholarly potential and made the Map Room the national map library nor probably managed to travel to most of the world, too (HW video 1994).

So on 30 April 1951 Wallis was appointed Assistant Keeper class 2 at the BM's Map Room to work under her supervisor R. A. Skelton who had become Superintendent in succession to Lynam. Both Lynam and Skelton had begun their careers in the BM's Department of Printed Books as cataloguers: Wallis was appointed directly to the Map Room and was the first academic geographer on the staff. She learnt, however, to catalogue from Anna Simoni who was in charge of Dutch acquisitions from 1950 to 1981 (HW video 1994). At that time the BM Library, formally the Department of Printed Books, expected its curatorial staff not only to work on the collections, acquiring, cataloguing and advising readers and dealing with important curatorial matters such as exhibitions, cartographical issues of diplomatic or of other national importance, but also to pursue appropriate scholarly work both in working hours and in their own time. The ideal therefore was a scholarly librarian or museum curator.

When Wallis joined the Map Room she became part of a long-standing scholarly tradition in the history of maps, discovery and exploration, which was in keeping with her own interests. Lynam, Skelton and their predecessors from the mid-nineteenth century had developed considerable knowledge in the history of cartography, discovery and exploration, as well as the necessary professional skills in library and cartographical work. Wallis soon followed in their footsteps.

All was not work, however, and she found time to belong to one of the Coffee Pot clubs for young professionals as well as the Fabian Society and the church (BL HWA, correspondence with John Beaglehole 1957). She was also a staunch member of the BM's choral society throughout her career. Early in her career she experienced a period of low self-esteem but managed, with counselling, to recover (St Hugh's 1997, 14; 2018 Saunders pers. comm.). Wallis also began to pay more attention to her appearance and to wear more becoming clothes instead of mirroring the dark suits of her male colleagues; this is evident from a comparison of the photos of her in the 1950s and those of her in her later career. She remained unmarried, although in 1961 she was engaged to be married to Ian Christie, a historian at University College London, but this fell through (Saunders, pers. comm. 2018) When on Saturday 8 July 1961 she told her colleague and friend F. G. Emmison, Essex County archivist, at a dinner of the Anglo–American conference, that the day was to have been her wedding day, he took her off to the 'Edmundo Ross club to cheer her up' (1995, 16 May letter from F. G. Emmison to author).

At that time there were few other women in the history of cartography field and in map librarianship and Wallis struck long-lasting friendships with many of them. In England there was the great historical geographer and historian of science and navigation professor Eva Germaine Rimington Taylor (1879–1966) whom Wallis knew and admired (see *Geographers* 31). Eila Campbell (1915–1994), professor of geography at Birkbeck College, University of London (1970–81); honorary secretary of the Hakluyt Society for many years; editor of the international journal of the history of cartography *Imago Mundi* for 20 years; and council member of the RGS and the Society for Nautical Research (*Imago Mundi* 1995, 47:1, 7–12), was a good colleague and friend and she and Wallis were often to be seen together, especially on conference trips, mulling over professional issues of the day while examining a globe or map. In France Myriam Foncin, first woman head of the Département des cartes et plans in the Bibliothèque nationale (1942–1963) and RGS Gill medallist (1961) was a friend and colleague as was Monique Pelletier, also head of the Département des cartes et plans from 1976. The two collaborated, among other things, on the ICA Historical Commission and in researching the globe maker for Louis XIV, Vincenzo Maria Coronelli. Another good friend and map curator was Ann-Mari Mickwitz of the Nordenskiöld collection in the University Library, Helsinki. In the United States there were her colleagues Jeanette Black, curator of the John Carter Brown Map collections, and Clara Le Gear, of the Maps Division of the Library of Congress; both Le Gear and Wallis were honoured for their respective contributions at the Society of Woman Geographers, Washington, DC, in 1994 (pers. know.). Wallis, educated at girls' school and college, always valued the friendships with these women in the library world, which was dominated by men.

In England Wallis was the first woman to be appointed at the level of deputy keeper in the Department of Printed Books as Superintendent of the Map Room in succession to Skelton in 1967. Wallis remembered that, in about 1964 (pers. comm.), Skelton had taken against her and later written a poor assessment of her when she went for promotion as his successor in 1967. This was a difficult time and one that dragged on before she was nonetheless appointed to the post (BL HWA, Letter to Brian Harley, 12 July 1967). As the BM's deputy director Maysie Webb later recalled, the appointment board decided in Wallis's favour because it was in the 'BM's interests and those of fairness'. The board was, of course, proved right (Tyacke and Campbell 1995, 18–19). The circumstances of her appointment and the pressure of being the first woman deputy keeper meant she was determined to do as well as, if not better than, those who had gone before. She always said that women had to be twice as good to be thought half as good as men. She knew the quotation from Charlotte Whitton, first woman mayor of Ottawa, which continued 'luckily this is not difficult' (pers. comm.). I do not think, however, she gave such discriminatory matters much serious thought, since for her what she did was so engrossing that she just brushed aside any personal feelings she may have had. As we have seen she was already set on a path of scholarship and the discovery and explanation of rare or little-known artefacts – in her case, maps and globes.

Maps and globes or the discovery of material culture

Wallis's first discovery came in 1949 when she found at Petworth House, West Sussex, the first edition (1592) of England's earliest terrestrial globe made by Emery Molyneux (*d.*1598), a maker of globes and ordnance, which showed the circumnavigations of Francis Drake from 1577 to 1580 and Thomas Cavendish from 1585 to 1588. A later edition of both terrestrial and celestial globes, dating from 1603, is in the Middle Temple in London.

The globes were financed by the London merchant William Sanderson, who presented them to Elizabeth I. Wallis's researches, and were published in the *Geographical Journal* in the 1950s and in 1987 when Anna Maria Crinò discovered that Molyneux had sailed with Drake, as confirmed by a contemporary manuscript account of Molyneux's workshop by Petruccio Ulbaldini (Wallis 1951, 275–90, 1952a, 89–90, b, 78, 1953a, 68–9; Wallis and Crinò 1987c, 11–18). The significance of the globe has resonated ever since in the literature and signifies Wallis's enduring scholarship in the then nascent history of globes in England and Europe. The late David Woodward appreciated Helen's considerable contribution to the history of globes: their

Figure 4.3 Helen Wallis examining the Molyneux terrestrial globe in 1953 (photo: BL HWA, 1953).

making, iconography and their use by contemporaries and then in education, especially in that of women and children in the eighteenth century (St Hugh's 1997, 35–43).

Wallis had a holistic approach to the history of cartography and included the use of globes and maps by all sorts of people, as part of their history. This approach is now common in historical and geographical studies, but was innovative in the 1950s when a positivist approach treated the history of cartography as one of continual improvement in geographical accuracy, while societal and cultural aspects were often excluded as irrelevant. Wallis recounted the story as it happened. This is particularly evident in her essay in the *Compleat Map Maker* (Thrower 1978b, 1–43), 'Geographie is better than divinitie', where the map selling and reading milieu of seventeenth-century London is brought to life through consideration of the maps and globes, especially those of the globe maker and map seller Joseph Moxon (1627–1691).

In 1962 Sir Percival David presented the BM with the earliest surviving Chinese globe dated 1623. This globe, once given its own seat on a plane as 'Mr Globe' when Wallis took it abroad for display, was painted in lacquer on wood and was made by the Jesuit fathers Manuel Dias the younger (Yang Manuo) and Nicolo Longobardi (Long Huamin). Wallis examined and described it in 1962 (with E.D. Grinstead 1962a). Joseph Needham and she corresponded about the significance of the globe and previous globes recorded in China but by then lost (BL HWA, letters Needham to Wallis 1963). Wallis continued with her researches into Chinese maps especially writing on the Matteo Ricci maps (1965c). With Howard Nelson and Yolande Jones (later Hodson) she put on an exhibition of Chinese and Japanese maps in 1974 (1974a), which reflected her long-standing interest and, of course, the great oriental collections of the British Library.

Wallis was a valued and active member of the Internationale Coronelli-Gesellschaft für Globenkunde (International Society for the History of Globes) founded in Vienna in 1952, publishing articles in their journal *Der Globusfreund* (1962b, 1967b, 1970b). This led inexorably to her work on the globes of Vincenzo Coronelli who by 1690 had assumed the role of the leading globe maker in Europe. He also produced atlases, among them his 1697 *Libro dei Globi,* which was an atlas of globes. In 1969 Wallis introduced readers to a facsimile of this great work (1969a) (and continued to publish on his globes (1971b, 1980a), especially those constructed for Louis XIV of 15 feet in diameter, which he completed in 1683 (now on display in Versailles)). Wallis's globe catalogue grew in density and it was eventually issued after her death as a *British Globes up to 1850: A Provisional Inventory* (1999) edited by Richard Dunn, who took on the work supervised by a working group chaired by the then Map Librarian Tony Campbell.

Map room to national map library in the wider world

If globes were an enduring interest, it was Wallis's work as Superintendent from 1967 and then Map Librarian that galvanized her energies into making the BM's Map Room, from 1973 the BL Map Library, into the national map library of the country. Her successor Tony Campbell considers that 'for some people learned societies are useful, both socially and academically … but for Helen the cat's cradle of interlocking personal and institutional relationships provided a frame work for almost everything she did' (St Hugh's 1997, 7–17). While furthering the aims of the Map Library, this network enabled Wallis's own collaborative researches into maps and globes, which always required other experts to contribute and to do so willingly for long periods of time. Her younger colleagues in the map room were fortunate to learn their own skills by doing so. Sometimes the deadlines for Wallis's multifarious tasks piled up a little too close for comfort, and she would be heard calling colleagues to 'action stations' – the usual sort of parlance among her nautical colleagues on the Society for Nautical Research council, and, of course, in countless wartime films. She also enjoyed a close collaboration with Mary Alice Lowenthal as her research assistant and editor from 1979 onwards, including in her retirement. Wallis always gathered together a strong team to realize whatever project she undertook.

Of especial note, as an example of her collaboration and support to others, is Wallis's essay on the 'Map collections of the British Museum Library' in *My head is a Map* in 1973, (1973b) celebrating the career of R. V. Tooley (1898–1986), the doyen of the London antiquarian map world. Tooley's production of the Map Collectors' Circle series of map studies had done so much to engage both academics and the wider interested public in early map studies. This same interest in bringing the history of mapping to a variety of audiences led Wallis to become a founder member of the International Map Collectors' Society (IMCoS) in 1980 and then its president in 1986. The society encompassed anyone interested in old maps: collectors, dealers, academics, librarians and map curators. On her death in 1995, the society generously renamed its annual award (previously the R.V. Tooley award) the Helen Wallis award in her honour. When she retired she continued to lecture and publish, as well as becoming an associate editor of another mapping journal *The Map Collector.*

In Wallis's time the Map Library staff grew from seven to 22 to manage the extensive collections of modern mapping and the introduction of machine readable cataloguing, as it was then called, as well as new staff for the reading room. Plans had been made for the new larger Map Library in the new BL building at St Pancras whose architects, Colin St John Wilson and M. J. Long, became well known to Wallis and colleagues, as they drew up their plans for the building in the 1970s. Wallis looked forward to seeing it operational (HW video 1994).

One of the most important parts of the Map Librarian's duties was to seek out and acquire maps that were of value to the collections and to scholars. Wallis had already proved able in this task before she became head, but under her leadership numerous important maps and atlases were bought for the collections. In particular, the library acquired the unique surviving example of Gastaldi's woodcut map of 1561 on nine sheets (St Hugh's 1997, 7–17) and the Royal United Services Institute (RUSI) map collection, which she and Bill Cummings unearthed on one of their forays into the map collections in England. The RUSI collection had been built up since the institution's foundation in 1831 and included three major historical collections, one of which was particularly important as it complemented the quality of those maps in the King George III collection that had been presented to the museum in 1823. This was the Amherst collection of maps, being the working maps of Field Marshal Jeffrey Amherst, who was commander-in-chief of British forces in North America from 1758 to 1763 (1970c, 39–40).

It was expected that as Map Librarian Wallis would be consulted on critical matters to do with the authenticity and reliability of maps. On such occasions she was fortunate to have the advice of Arthur Baynes-Cope of the BM's Research Laboratory. He was an expert in the field of document examination and he and Helen shared a love of globes. He is best remembered for his pioneering work in *The Study and Conservation of Globes* (1985). Wallis and Baynes-Cope took a long-standing interest in the famous, perhaps notorious, case of the Vinland map. Wallis's own interest stemmed from her professional interest in the map's authenticity, and because, in particular, the map had been examined in the 1960s, if not before, in the BM, before being purchased by Yale.

Wallis's predecessor, Skelton, was the author, with Thomas E. Marston and George Painter, of the *Vinland Map and Tartar Relation* (1965). This map, apparently drawn in the mid-fifteenth century and showing Norse discoveries in North America, remains a controversial artefact on a number of grounds, not least because, if it is authentic, it would be a map showing the discovery of North America well before Columbus in 1492. Throughout her career Wallis remained sceptical about its authenticity. She encouraged those who researched the provenance and debatable content of the map and in 1974 co-organized a conference at the RGS, the proceedings of which were published as 'The strange case of the Vinland map: A symposium' (1974c, e). Yale University, the owners of the map, had asked the firm of analytical chemists McCrones (McCrone 1999) of Chicago to examine the map and at the symposium they revealed that the map's ink contained anastase, a substance used only after about 1923.

The debate, heated in some quarters, continues to this day among chemists, between chemists and other analytical scientists, and among historians of varying sorts, who have more or less expertise in the elements necessary to examine the map itself: for example, the difficulty of being assured of the map's provenance and the apparent modern outline of Greenland. Wallis kept abreast of all the developments and in particular advised Kirsten Seaver in her intriguing and meticulous detective work about the map's genesis: *Maps, Myths and Men: The Story of the Vinland Map* (2004). Seaver posed the question: If the map was not a fifteenth-century one, then who drew it in the twentieth century? She shows that a likely candidate could be the Jesuit scholar Josef Fischer (1858–1944) known for his works on fifteenth-century cartography, though this interpretation has recently been challenged (Harvey 2006). Helen's role in this long-running saga was that of a mapping adviser who also knew much about its recent history from the 1960s onwards (1990c, 1991f; HW video 1994).

National and international organizations

The organizations and societies in which Wallis was involved benefitted from her commitment to the map and cognate worlds, sometimes owing their foundation to her or having her as president or chair of a committee. Examples include the British Cartographic Society (BCS); the Map Curators' Group (MCG) of the BCS (*est.* 1967); the Geography and Maps Section of the International Federation of Library Associations (IFLA) founded by her with Walter Ristow of the Library of Congress; or the History of Cartography working group, later a Commission of the International Cartographic Association (ICA) in 1972 (1979c, d, e); chairmanship of the Society for Nautical Research 1971–1988; the councils of the RGS, the Hakluyt Society, and many others. Her great networking strength, however, ate into the time she had for her own scholarship since she rarely, if ever, declined any invitation to collaborate with others in some map-related venture or to lecture on maps anywhere at home or abroad.

An important contribution to the work of the wider cartographic community was Wallis's co-editing with Arthur Robinson, professor emeritus of cartography at the University of Wisconsin–Madison, of *Cartographical Innovations: An International Handbook of Mapping Terms to 1900*, published by the ICA in conjunction with Map Collector publications (1987a). Wallis had already worked for many years on a range of academic mapping subjects, such as the emergence of statistical and thematic mapping. As early as the late 1960s she and Sarah Jeacock (later Tyacke), put up an exhibition on statistical maps entitled 'Uses of Statistics' (BM 1 September 1969–17 January 1970). Arthur Robinson was instrumental in her researches and they co-authored 'Humboldt's map of isothermal lines: a milestone in thematic cartography' for *Cartographic Journal* (1967a). Wallis continued to provide critical information to Robinson for his work on *Early Thematic Mapping in the History of Cartography* (1982).

Cartographic Innovations proved to be a useful volume of reference for all historians of cartography. However, by the time the book, which had been begun a long time earlier and had more than 100 contributors, came out in 1987, a new approach was sweeping through the history of cartography. Developed in the 1980s by J. Brian Harley and David Woodward, it bore fruit in their mammoth History of Cartography – a publication project covering the history of cartography in all parts of the world from prehistoric times to the end of the twentieth century. Harley in particular had critical views about *Innovations,* not about its accuracy and practical use, which were self-evident, but about its assumptions. In a review article 'Innovation, social context and the history of cartography', he described the volume as 'determined in its approach to provide a positivist and linear view of cartographic history in which "innovations" punctuate the linear "advances" of technical progress' (Harley 1987, 59 in *Cartographica* 24, 4 (Winter), 59–68). This approach ran counter to his own, which explored cartography as part of the society in which it existed. The ICA's approach, as embodied in *Innovations,* inevitably fell foul of this new history of cartography, which, in its turn, did not go unopposed. With the benefit of hindsight, Mathew Edney, now the general editor of the *History of Cartography*, categorized what he saw as the four different approaches to the history of cartography at the beginning of the twentieth century: 'traditional, with an emphasis on map content, and then internal map history on map form and the history of map making'. He then categorized the new 'Harley' turn in the 1980s as being towards a 'socio-cultural map history with an emphasis on map meaning (culture), function (society) and ideology' as well as having an 'emphasis on the processual map history with an emphasis on the process of production circulation and consumption' (2014, *Imago Mundi* 66 Supp., 83–105 quotation at 84).

Wallis preferred the careful description of maps as artefacts, together with a close examination of their content, their production, circulation and consumption, and a consideration of who their readers were at the time in question. While evidently traditional in her approach, she was also much more in line with the later sociocultural turn *avant la lettre,* as we have seen.

Wallis's final work for the wider historical and map community was the publication of the Royal Historical Society's *Historian's Guide to Early British Maps* (1995). Wallis had begun the project in the 1970s with the help of colleagues in the Map Room, following Skelton's initial proposal, but it stalled until her retirement when she took it up again. It was co-edited in the end by Anita McConnell (see *Geographers* 36) as Wallis's health no longer allowed her to direct the work, though McConnell characteristically acknowledged Wallis's contribution and seniority in the field by naming her as first editor.

Exhibitions and publication collaborations

In her variety of approaches and subjects Wallis's interests coincided with those of many researchers within and outside the museum and library world, and she was able to persuade them to collaborate across a wide range of map-related endeavours. This is evident in her prolific publishing output (see her selected bibliography below) and was also evident in her contribution to BM and BL exhibitions. The exhibitions created, as well as reflected, interest in maps and their history especially in the United States and Australia, and made Wallis probably the best-known historian of cartography in the English-speaking world. This, in turn, redounded to the reputation of the BL's Map Library.

Wallis's most celebrated exhibitions were those that harnessed the collections and curators of a number of BM and BL departments to reveal the riches of their collections and to elucidate for visitors to the BM's King's Library the stories of these events. They were based on her own long-standing research interests. These exhibitions, which later toured venues in the United States, were 'The American war of Independence 1775–83' (1975a); 'Sir Francis Drake: an exhibition to commemorate Francis Drake's voyage round the world 1577–1580' (1977d); and 'Raleigh and Roanoke: the first English colony in America' (1985e), which explored Raleigh's 1585 colony in present day North Carolina, the first English attempt at a permanent settlement in North America. Then there were the two exhibitions celebrating Cook's voyages of 1768–1771, 1772–1775 and 1776–79. To the first, 'Captain Cook's first voyage round the world (19 July–27 October 1968): an exhibition to commemorate the bicentenary of first voyage of Captain Cook' (BM, 1968), Wallis contributed a major part; it then went to Australia. To the second exhibition, 'Captain Cook in the South Seas 15 February 1979–29 September 1980', curated by Hugh Cobbe of the Department of Manuscripts, Wallis contributed her expertise (1979a). She continued to write on subjects relating to Cook and the Pacific to the end of her life (1979f, 1994b). On one of her trips to Australia in March 1979 she accompanied the exhibition artefacts, which were in five huge crates weighing apparently six tons (1970a). It would be difficult not to admire the immensity of scope, the number of artefacts chosen and the panache of these celebratory exhibitions from 1968 onwards. They played to Wallis's strengths and long-standing knowledge of the Pacific, from her doctoral dissertation of 1954 through many other articles (1961a, 1963, 1964, 1978d, 1986g, 1994b). At the suggestion of Skelton and encouraged by John Beaglehole, editor of Cook's voyages, Wallis meticulously edited Philip Carteret's *Round the world voyage 1766–1769* (1965a), providing an excellent account of the genesis of the voyage, demonstrating what was known beforehand and what was accomplished. Wallis's cartographic knowledge was, as usual, to the fore, for example, in explaining the mysteries of the whereabouts of the Solomon Islands. On 2 November 1962, Beaglehole inscribed for Wallis on the flyleaf of a copy of the first volume of his *The Journals of Captain Cook: The Voyage of the Endeavour 1768–1771* (Hakluyt Society, extra series, 34) 'The Martyr to Cook … for the Martyr to Carteret' (priv.coll.).

Andrew David, a hydrographical expert and Cook scholar, in commenting on Wallis's researches and publications on the Pacific (*St Hugh's* 1997b, 22–30) notes in particular her attendance at a Pacific conference in 1978 at Simon Fraser University in Vancouver to mark the bicentenary of Cook's exploration

Figure 4.4 Helen Wallis on her travels at Uluru (Ayers Rock) Australia in 1988 (photo: priv. coll).

of the north-west coast of America during his last voyage. Wallis's paper on 'The publication of Cook's journals: some new sources and assessments' dealt with the charts as well the journals, charts on which David subsequently relied in *The Charts and Coastal Views of Captain Cook's Voyages* (3 vols. 1988–1997). David drew Wallis's attention to the running journal kept by James King, which covered the period from the expedition's last visit to Petropvlosk in Kamchatka to the ship's arrival at the Cape of Good Hope and which he had discovered in the Hydrographic Office in London. Wallis was hoping to transcribe and edit this in retirement for the Hakluyt Society and had begun the work, but her death intervened. In the event in 1997, David included it, in her memory, in volume 3 of *Charts and Views* (1997, 'James King's Running Journal' 271–310). A further Cook article that she and her BL colleague David Paisey were working on at the time of her death was finally finished by Paisey in 2011 as 'Letters by Johann Reinhold Forster about Captain Cook's first voyage and the preparations for the second. Translated from German, with an introduction and notes, by David Paisey' (*Terrae Incognitae* 43,2, 99–133).

As is evident Wallis advised and collaborated generously with scholars and researchers in their own endeavours. Thus she worked with Beaglehole on the journals of Captain James Cook's voyages and David Quinn on *The Hakluyt Handbook* in which she also wrote 'Edward Wright and the 1599 world map' (1974g) and contributed the chapter on Hakluyt and the Pacific where she demonstrated that for Hakluyt the Pacific was mainly important as a route to the Orient (1974h). She also worked with Quinn on the history of the Roanoke colony, participating with her usual enthusiasm, at a conference there in 1993. She collaborated with many other researchers as is shown by her bibliography (see below).

Jean Rotz *Boke of Idrography* 1542 presented to Henry VIII

Wallis probably considered her most important publication to be her edited edition of *The Maps and Text of the Boke of Idrography Presented by Jean Rotz to Henry VIII, Now in the British Library* (1981f). The edition was made for Viscount Eccles, chairman of the BL, with a foreword by Eccles for the Roxburghe Club. It exemplifies the range of her international scholarly network and the culmination of her own knowledge and researches. She drew on her extensive knowledge of the royal map collections to place the atlas in its Royal Library context and on her extensive knowledge of the early Pacific voyages, augmented by the researches of her French colleagues on the work of the Dieppe school of cartography, to demonstrate Rotz's cartographic and navigational achievement. To produce the commentaries necessary for the atlas facsimile, which needed to be mathematical and navigational as well as geographical and historical, she enlisted Janet Backhouse of the Department of Manuscripts, who wrote on Rotz and his contemporaries at Henry VIII's court; David Quinn, who wrote on the Americas; and Paul Hair, who wrote on Africa. Derek Howse and David Waters of the National Maritime Museum advised on astronomy and navigation, and Charles Beckingham and Michael Rogers of the BM advised on Asia. Wilma George (then at Lady Margaret Hall, Oxford) was a good friend of Wallis and advised on Rotz's drawings of animals. Mary Alice Lowenthal prepared the bibliography and helped to see the volume through Oxford University Press. In France, Monique Pelletier and Roger Hervé provided much material and advice and Abbé Roger Desremaux, Professor of Geography at the University of Lille, researched the provincial archives for Rotz and his family. Gathering all the contributions together was a vast editorial operation, which resulted in a beautiful production of the highest scholarly standard.

Wallis's own contributions were to explain Rotz's treatise on navigation 'The differential Quadrant and the triacte' and to assess Rotz's claims for the *Boke*, and to place him as a cartographer in the context of the Dieppe school of hydrography. This led her to consider Rotz's depiction of a large land mass south of Java and the Dieppe School's cartographic tradition of showing this as 'Java la Grande' and thus its possible identification as Australia. Could it have been discovered by or known to the Portuguese who frequented the East Indies before the Dutch, and whose charts the Dieppe school used in their own compilations? This possibility led her to title her contribution 'The works of Jean Rotz: Java la Grande, the first sight of Australia'.

The hypothesis about Australia intrigued Wallis for the rest of her life although other commentators, in particular Bill Richardson of Flinders University, dispute this hypothesis (*The Portuguese Discovery of Australia* (2016)). She was always close to where a discovery or revelation might be and pursued researches as far as she could from extant sources and maps – no more so than in one of her last studies: that of the Rawlinson copperplate maps in the University of Oxford's Bodleian Library, which she identified with Pearce S. Grove in 'Discovery of the Rawlinson copper plate maps of the Americas and their related prints' (1991a). The prints were hitherto unknown, but she found them at Holkham Hall in Norfolk in John Innys's *System of Cosmography*, completed about 1750, prints from all the Rawlinson plates in the Bodleian, except for one item that remained to be found. She then described how she and Mary Alice Lowenthal identified the missing print of Boston (1993b and c). This was her last foray into the discovery of maps begun in 1949 with the Molyneux globe at Petworth House.

Conclusion

Helen Wallis's field of scholarship is part of what is now called material culture and, unlike in the 1950s when Wallis began her career, maps and globes are now considered relevant to many types of historical and geographical enquiry, as well as being the object of research in their own right. There had been for a very long time, and continues to be, interest in early maps by antiquarians and collectors, but few modern cartographers were or are interested in their profession's origin or in early mapping. There were few practitioners and even fewer books on the history of cartography. Map librarianship was hardly considered by book librarians. There was therefore room for a considerable and comprehensive change and Wallis was a very visible catalyst for that change. The current organizational, professional framework of the map world owes a great deal to her as a prime mover in setting up new, or actively supporting, map history organizations, but always in collaboration with other like-minded colleagues.

Wallis was appointed Officer of the Order of the British Empire in 1986 on her retirement for her services as Map Librarian of the BL, (Supplement to the *London Gazette*, 14 June 1986). She was made Hon DLitt of Davidson College, North Carolina, in 1995 and received an honorary fellowship of Portsmouth Polytechnic (now University) in 1988, and was awarded the Caird Medal of the National Maritime Museum (1990) (Mead 1995a and b).

Wallis served on the councils of the Society of Antiquaries of London and the RGS, which awarded her its prestigious Victoria medal after her death. Having held the presidency of the British Cartographic Society and co-founded its map curators' group, she was awarded its silver medal in 1988. Her fellow librarians had earlier honoured her with an honorary fellowship of the Library Association in 1985. She retired in 1987 (Mead 1995a and b). She was a Fellow of the Society of Antiquaries of London and of the Royal Historical Society (Howse and Gueritz 1995). Her significant contributions to scholarship were celebrated at the RGS's commemoration of her life and work on 9 May 1995. A record of the tributes given there and others from her friends and colleagues who could not be present was published as *The Globe My World: Tributes to Helen Wallis,* edited by Sarah Tyacke, her former deputy, and her successor Tony Campbell. Over sixty tributes were printed by Dr Robin Alston, a long-standing BL colleague, and distributed to colleagues and friends (Tyacke and Campbell 1995). Wallis deposited her papers in the BL Map Library where colleagues named the archive 'Mount St Helen's' much to her amusement and delight. She remained in her London flat and continued to write and publish. She kept up her lifelong interest in music and towards the very end of her life she could be found playing the xylophone for the patients at her local hospice.

Wallis's health began to fail after she returned from Australia in 1994. She began to make preparations for the time when she could no longer work, and enlisted the help of colleagues to finish incomplete projects. She died of cancer on 7 February 1995 at the Hospital of St John and St Elizabeth in St John's Wood, London. In a visit shortly before her death, Mead (1995) noted her continued optimism and quiet laughter in the hospice, which reflected her Christian beliefs. Her funeral service was held on 10 February at St John's Wood Church where she worshipped, and was followed by many obituaries from all the fields in which Helen Wallis was active (Mead 1995a and b; Campbell 1995; Tyacke1995; Scott and Campbell 1995). A memorial seminar was held in her honour at St Hugh's College, Oxford, on 12 April 1997, a record of which was issued to friends and others. She was memorialized in the *Oxford Dictionary of National Biography* (Campbell 2004). Her extensive library of books on the history of discovery, exploration and, of course, cartography were donated to the Osher Library in Portland, Maine, United States. Her great friend John Maggs of the famous London booksellers Maggs Brothers arranged for the donation, which was finally delivered on 14 February 1996 (Osher Library, pers. comm., Edney 2018).

Wallis's academic strengths were those of a scholar committed to her own fields of the histories of cartography, discoveries and voyages. She demonstrated this through her own research and publications; by organizing exhibitions; by augmenting the map collections of the BL; and through giving advice and explaining to countless audiences, at home and abroad, aspects of the history of cartography, travel, exploration and discovery. That such promotion of the public understanding of the history of cartography is now required practice for all collections, is in large measure due to the pioneering work of Wallis and like-minded colleagues. In Wallis's case it was based on deep personal knowledge gathered during more than forty years in the map world. Such was the goodwill generated by her in her career, those projects that she had left undone on her death were carried on for her by her colleagues. Even more telling is the establishment of the Helen Wallis Fellowship 1995 in her memory at the BL for researchers to consult the library's cartographic collections, in particular to explore the interdependence of cartographic and other sources in historical investigation that have an international dimension (BL Map Library proposal for the Helen Wallis Fellowship (1995 priv.coll.)).

Acknowledgements

My thanks to Helen's successors Tony Campbell and Peter Barber both of whom, like me, worked with Helen and who gave me advice and help in writing this essay. I would also like to thank Tom Harper, the present lead curator for antiquarian maps in the British Library, especially for helping me with the Helen Wallis archive. My thanks also go to Helen's long-standing friend Ann Saunders whom I visited and to Mary Alice Lowenthal, Helen's good friend, research assistant and editor from 1979 onwards. The views are my own when considering her place in the history of her profession and academically and they also inevitably reflect my period of working with Helen for over 20 years in the BM and BL and as her literary executor.

Selected bibliographical works by Helen Wallis

This bibliography is based on the work of Terry Kay, of the BL, who compiled a complete bibliography of Helen Wallis's published works on the occasion of her retirement from the library. The bibliography was published in *The Map Collector* (40, Autumn 1987, 30–8). It follows the selection for the bibliography included by Tony Campbell in his obituary in *Imago Mundi* 1995 I 47 199–192. Only major articles, review articles and longer prefaces are included; for shorter reviews and prefaces, reports and obituaries written before autumn 1987, Kay's bibliography in *The Map Collector* should be consulted. Those publications completed by others after 1996 are referred to in the text.

1951 'The first English globe: A recent discovery', *Geographical Journal* 117, 275–90.

1952a 'The Molyneux globes', *British Museum Quarterly* 16, 89–90.

1952b 'A newly-discovered Molyneux globe', *Imago Mundi* 9, 78.

1953a 'A contemporary handbook to the Molyneux globes', *British Museum Quarterly* 18, 68–9.

1954a 'The exploration of the South Sea, 1519 to 1644', University of Oxford unpublished doctoral dissertation, 3 vols and portfolio of maps at BL Maps 186.m.1.

1955b 'Further light on the Molyneux globes', *Geographical Journal* 121, 304–11.

1961a 'Abel Janszoon Tasman', *Encyclopædia Britannica* (ed. in chief Harry S. Ashworth). London: Encyclopædia Britannica Ltd., 21, 826.

1962a (with E. D. Grinstead) 'A Chinese terrestrial globe, A.D.1623', *British Museum Quarterly* 25, 83–91.

1962b 'The first English terrestrial globe', *Der Globusfreund* 11, 158 [German translation, 159].

1962c 'Globes in England up to 1660', *Geographical Magazine* 35, 67–79.

1963 'John Hawkesworth and the English circumnavigators', *Commonwealth Journal* 4, 167–71.

1964 'The Patagonian giants', in Robert E. Gallagher (ed.), *Byron's Journal of his Circumnavigation 1764–1766*. Hakluyt Society, second series, 122; Cambridge, Cambridge University Press for the Hakluyt Society, appendix 1, 185–99.

1965a (as editor) *Carteret's Voyage Round the World 1766–1769*, 2 vols. Hakluyt Society, second series, 124 and 125. Cambridge: Cambridge University Press for the Hakluyt Society.

1965b 'English enterprise in the region of the Strait of Magellan', in J. Parker (ed.), *Merchants and Scholars: Essays in the History of Exploration and Trade Collected in Memory of James Ford Bell*. Minneapolis: University of Minnesota Press, 193–220.

1965c 'The influence of Father Ricci on Far Eastern cartography', *Imago Mundi* 19, 38–15.

1965d 'Landmarks in atlas cartography', *Cartographic Journal* 2(1), 42–14. [Review article of *Abraham Ortelius: Theatrum Orbis Terrarum* 1570, facsimile ed. with an introduction by C. Koeman; and of *Lucas Jansz. Waghenaer: De Spieghel der Zeevaerdt* 1584–85, facsimile ed. with an introduction by C. Koeman].

1965e 'Report on the library classification of books and maps', *Cartographic Journal* 2(1), 14–15.

1966 'The rôle of a national map library', *Cartographic Journal* 3(1), 11–14.

1967a (with A. H. Robinson) 'Humboldt's map of isothermal lines: A milestone in thematic cartography', *Cartographic Journal* 4(2), 119.

1967b 'Thematische Elemente auf frühen englischen Globen (1592–1900)', *Der Globusfreund* 15–16, 75–85.

1968 'The use of terrestrial and celestial globes in England', *Actes du Xle Congrès International d'Histoire des Sciences* (Breslau, Ossilneum Maison d'Edition de l'Académie Polonaise des Sciences) 4, 204–12.

1969a Bibliographical note, in *Vincenzo Coronelli; Libro dei Globi* (*Venice 1693, 1701*). Theatrum Orbis Terrarum, 4th series, vol. 5; Amsterdam, Theatrum Orbis Terrarum, v–xxii.

1969b Preface to George, W., *Animals and Maps*. London, Secker and Warburg, 7–19.

1969c 'Reference work in a map library', *Inspel* 4, 85–8.

1969d Text concerning maps, in Shigek Goto (ed.), *The British Museum*, The Great Museums of the World 8. Tokyo: Ghogakukan, 175–6 (printed with Japanese text but with an English title as given above).

1970a 'Captain Cook and the rediscovery of Australia', Currency (monthly staff bulletin).

1970b 'Coronelli's Libro dei Globi', *Der Globusfreund* 13–20, 124–9.

1970c (with S. Jeacock) 'Royal United Services Institution Map Collection', *Cartographic Journal* 7(1), 39–40.

1971a (with W. P. Cumming) 'Charts by John Friend preserved at Chatsworth House, Derbyshire, England', *Imago Mundi* 25, 81.

1971b 'The globes of Vincenzo Maria Coronelli', in *Xlle Congrès International d'Histoire des Sciences, Paris 1968. Actes*. Tome Xa: *Histoire des Instruments Scientifiques*. Paris: Albert Blanchard, 107–13.

1971c 'Reference work in a map library', in L. Zögner (ed.), *Karten in Bibliotheken: Festgabe für Heinrich Kramm zur Vollendung seines 65. Lebensjahres*, Kartensammlung und Kartendokumentation 9. Bonn–Bad Godesberg: Bundesforschungsanstalt für Landeskunde und Raumordnung Selbstverlag, 111–16.

1972a 'England re-discovered', *British Museum Society Bulletin* 9, 20.

1972b 'Richard Hakluyt', in C. C. Gillispie (ed.), *Dictionary of Scientific Biography* 6. New York: Charles Scribner for the American Council of Learned Societies, 20–1.

1972c 'Introduction', in *Atlas of the British Isles: By Pieter van den Keere c. 1605*. Lympne Castle, Kent: Harry Margary, iiv.

1972d (with S. Tyacke) 'Progress in the map world', *New Library World* 73, 295–6.

1973a 'Canadian maps in the map collections of the British Museum', in *Proceedings of the Association of Canadian Map Libraries*. Ottawa: Association of Canadian Map Libraries, 84–94.

1973b 'The map collections of the British Museum library', in H. Wallis and S. Tyacke (eds), *My Head is a Map: Essays and Memoirs in Honour of R. V. Tooley*, 3–20. London: Francis Edward and Carta Press. (The preface (xi–xvi) to this work, by H. Wallis, was also published in the *Antiquarian Booksellers' Association Miscellany* 15 (1974): 33–7.)

1973c 'Maps as a medium of scientific communication', in Jôzef Babicz (ed.), *Studia z dziejaw geografii i kartografii*, Monografie z dziejöw nauki i techniki 87. Breslau: Polska Akademia Nauk, 251–62.

1974a (compiler with H. Nelson and Y. Jones) *Chinese and Japanese Maps*. London: British Museum Publications Ltd. for the British Library Board [exhibition catalogue].

1974b 'Introduction', in *Album Antoine de Smet*. Brussels: Centre National d'Historié des Sciences, xix–xxiii.

1974c Introduction to 'The strange case of the Vinland Map: A symposium', *Geographical Journal* 140, 183–6.

1974d 'Abel Janszoon Tasman', in *The New Encyclopaedia Britannica*, Macropaedia. Chicago: Encyclopaedia Britannica Inc., 17, 1070–1. [Revised version of the article first published in the *Encyclopaedia Britannica*, 1961.]

1974e 'The Vinland map', *Cartographic Journal* 11(1), 58.

1974f 'Working group on the history of cartography', *International Geographical Union Bulletin* 25(2), 62–4.

1974g 'Edward Wright and the 1599 world map', in D. B. Quinn (ed.), *The Hakluyt Handbook*. (Appendix to the article 'Hakluyt's maps' by R. A. Skelton), vol. 1. London: Hakluyt Society, 69–73.

1974h 'The Pacific', in D. B. Quinn (ed.), *The Hakluyt Handbook*, vol. 2. London: Hakluyt Society, 223–33.

1975a (with S. Tyacke, Y. O'Donoghue, H. Cobbe and R. Smith), *The American War of Independence 1775–83; A Commemorative Exhibition*. London: British Museum Publications Ltd for the British Library.

1975b 'Missionary cartographers to China', *Geographical Magazine* 47, 751–9.

1975c 'The training of map curators in Great Britain', *Inspel* 10, 28–33.

1976a Bibliographical notes to accompany the facsimile reproduction of *Symonson's Map of Kent, 1596: As Published by Stent about 1650*. Southampton: Ordnance Survey.

1976b (with P. K. Clark) 'Education and training of map librarians: Special needs of developing countries', *Inspel* 11, 100–2.

1976c 'From Dr Halley to Captain Cook: Some British contributions to the mapping and charting of Brasil, 1700 to 1800', *Anais Hidrográficos* 33(Suppl.), 129, 16.

1976d as editor *Map-Making to 1900: An Historical Glossary of Cartographic Innovations and Their Diffusion*. London: The Royal Society for the International Cartographic Association.

1977a Bibliographical notes to accompany the facsimile reproduction of *California and the Seven Cities of Civola: Chart of North-West America by Joan Martines 1578*. London: British Museum Publications Ltd. for the British Library.

1977b Bibliographical notes to accompany the facsimile reproduction of *Land of the Codfish: Chart of the North Atlantic by Joan Marlines 1578*. London: British Museum Publications Ltd. for the British Library.

1977c Bibliographical notes to accompany the facsimile reproduction of *The West Indies and the Spanish Main: Chart of the Gulf of Mexico and the North-East Coast of America by Joan Martines 1578*. London: British Museum Publications Ltd. for the British Library.

1977d (compiler with S. Tyacke and P. Higgins) *Sir Francis Drake: An Exhibition to Commemorate Francis Drake's Voyage Around the World* 1577–1580. London: British Museum Publications Ltd. for the British Library.

1977e 'Map libraries', in H. A. Whatley (ed.), *British Librarianship and Information Science 1971–1975*. London: Library Association, 246–51.

1977f 'Silver medal for the *Golden Hind*', *Geographical Magazine* 50, 112–17.

1978a 'Cartography in Great Britain in the eighteenth century: A survey of topographical mapping', in *La Cartographie au XVIIIe siècle et l'oeuvre du Comte de Ferraris (1726–1814): Colloque International Spa, 8–11 Sept 1976*. Bruxelles: Crédit Communal de Belgique, 165–77.

1978b 'Geographie is better than divinitie: Maps, globes and geography in the days of Samuel Pepys', in N. J. W. Thrower (ed.), *The Compleat Plattmaker: Essays on Chart, Map and Globe Making in England in the Seventeenth and Eighteenth Centuries*. Berkeley: University of California Press, 1–43.

1978c 'The place of globes in English education, 1600–1800', *Der Globusfreund* 25–7, 103–10.

1978d 'Publication of Cook's journals: Some new sources and assessments', *Pacific Studies* 1, 162–94.

1979a 'Conclusion', in H. Cobbe (ed.), *Cook's Voyages and Peoples of the Pacific*. London: British Museum Publications Ltd., 129–42.

1979b (with S. Tyacke) 'Department of Printed Books: Acquisitions 1968–1978 Map Library', *British Library Journal* 5, 181–97.

1979c 'Walter W. Ristow', in H. Wallis and L. Zögner (eds), *The Map Librarian in the Modern World: Essays in Honour of Walter W. Ristow*. Munich, K. G. Saur for the IFLA Section of Geography and Map Libraries, 11–19.

1979d 'Map librarianship comes of age', in H. Wallis and L. Zögner (eds), *The Map Librarian in the Modern World: Essays in Honour of Walter W. Ristow*. Munich, K. G. Saur for the IFLA Section of Geography and Map Libraries, 107–16.

1979e (ed. with L. Zögner) *The Map Librarian in the Modern World: Essays in Honour of Walter W. Ristow*. Munich, K. G. Saur for the IFLA Section of Geography and Map Libraries, 295.

1979f 'The voyages of Captain Cook', *Hydrographic Journal* 15, 7–13.

1979g *The Voyage of Sir Francis Drake Mapped in Silver and Gold*. The Friends of the Bancroft Library Keepsake 27; Berkeley: University of California.

1980a (with M. Pelletier) 'Les globes du Roi Soleil', in *Cartes et Figures de la Terre*. Paris: Centre Georges Pompidou, xii–xiv [exhibition catalogue].

1980b Preface, *The History of the Cartography of Cyprus*, by A. Stylianou and J. A. Stylianou. Nicosia: Cyprus Research Centre, xi–xii.

1980c (with M. Pelletier) 'The resurrection of Coronelli's great globes', *Map Collector* 13, 22–6.

1980d 'Some new light on early maps of North America, 1490–1560', in C. Koeman (ed.), *Land- und Seekarten in Mittelalter und in der frühen Neuzeit*, Wolfenbütteler-Forschungen 7. München: Kraus International Publications, 91–121.

1981a 'Early maps as historical and scientific documents', in K. Hakulinen and A. Peltonen (eds), *Papers of the Nordenskiold Seminar on the History of Cartography and the Maintenance of Cartographic Archives, Espoo (Finland), 12–15 Sept 1979*. Helsinki: Nordenskiöld-Samfundet, Finland, 101–16.

1981b 'The royal map collections of England', *Revista da* Universidade de Coimbra 28 (1980), 461–8.

1981c 'The history of land use mapping', *Cartographic Journal* 18(1), 45.

1981d 'Introduction to the geography and map libraries section of IFLA', *Inspel* 15, 108–13.

1981e 'New light on the Dieppe maps', in D. Howse (ed.), *Five Hundred Years of Nautical Science: Proceedings of the 3rd International Reunion for the History of Nautical Science and Hydrography held at the National Maritime Museum, Greenwich, 24–28 Sept 1979*. London: National Maritime Museum, 72–9.

1981f (as editor) *The Maps and Text of the Boke of Idrography Presented by Jean Rotz to Henry VIII now in the British Library.* Oxford: The Roxburghe Club.

1981g 'Cartographic innovation: An historical perspective', in *Canadian Institute of Surveying Centennial Convention Proceedings/Actes: Ottawa, Canada* 19–23 April, 1981. Ottawa: Canadian Institute of Surveying, 50–63.

1982a 'The Dieppe maps—the first representation of Australia?' *The Globe* (Journal of the Australian Map Curators' Circle) 17, 23–50.

1982b 'The enigma of Java-la-Grande', in *Australia and the European Imagination: Papers from a Conference Held at the Humanities Research Centre, Australian National University, May 1981*, Humanities Research Centre Monograph 1. Canberra: Australian National University, 1–40.

1982c 'Introduction', *Printed Maps of Sussex 1575–1900*, by D. Kingsley. Lewes: Sussex Record Society 72, ix–xxi.

1982d 'The Rotz atlas: A royal presentation', *Map Collector* 20, 40–2.

1982e 'Understanding maps', in *The Great Geographical Atlas*. London: Mitchell Beazley International Ltd., 98–103.

1983a 'British map collections: Cooperative planning and projects', *Inspel* 17, 239–50.

1983b 'Foreword', to Shirley, R. W., *The Mapping of the World: Early Printed Maps 1472–1700*. London: Holland Press, xi–xvi.

1983c Letters to the editor on the history of the Falkland Islands, *Geographical Journal* 149, 276–7, 400–1.

1984a 'A banquet of maps: An account of the map collections of the British Library', *Map Collector* 28, 2–10.

1984b 'The cartography of Drake's voyage', in N. J. W. Thrower (ed.), *Sir Francis Drake and the Famous Voyage, 1577–1580: Essays Commemorating the Quadricentennial of Drake's Circumnavigation of the Earth*. Berkeley: University of California Press, 121–63.

1984c 'Did the Portuguese discover Australia? The map evidence', *Technical Papers of the 12th International Conference at Perth, 6–13 Aug 1984*. Perth, 12th ICA Conference Committee 2, 203–19.

1984d 'England's search for the northern passages in the sixteenth and early seventeenth centuries', in L. Rey et al. (eds), *Unveiling the Arctic: Proceedings of the First International Conference on the Discovery and History of the Boreal Polar Regions*. Calgary: The Arctic Institute of North America/University of Alaska Press, 453–72. (Also published in *Arctic* 37(4), 453–72.)

1984e 'Raleigh and Roanoke: The beginnings of Anglo-America', The American Trust for the British Library, *Newsletter* 11, 1–2.

1984f 'Resources for advanced research in Canadian studies in the British Library Reference Division: Maps', in P. M. Turner (ed.), *Canadian Studies: Papers Presented at a Colloquium at the British Library 17–19 Aug 1983*, British Library Occasional Papers 1. London: British Library Reference Division Publications, 49–65.

1984g 'Resources for Australian and New Zealand studies in the British Library Reference Division: Maps', in P. M. Turner (ed.), *Australian and New Zealand Studies*: Papers *Presented at a Colloquium at the British Library 7–9 Feb 1984*, British Library Occasional Papers 4. London: British Library Reference Division Publications, 105–15.

1984h 'Terra Australis, Australia and New Zealand: Voyages, discoveries and concepts', in P. M. Turner (ed.), *Australian and New Zealand Studies: Papers Presented at a Colloquium at the British Library 7–9 Feb 1984*, British Library Occasional Papers 4. London: British Library Reference Division Publications, 184–93.

1984h 'Sir Walter and a legend of Indians', *British Museum Quarterly* 46, 10–13. (An apology for the omission of Helen Wallis's name as author of this article was published in *British Museum Quarterly*, No. 47.)

1985a (with A. H. Sijmons) *Atlas de Santarem: Facsimile of the Final Edition 1849: Explanatory Notes*. Amsterdam: Rudolf Müller (Booklet to accompany facsimile edition of *Atlas composé de mappemondes* … by the Viscount de Santarem, 3rd edn, 1849).

1985b 'Emigré map-makers and the protestant New World', *Proceedings of the Huguenot Society of London* 24, 210–20.

1985c 'Die Kartographie der Jesuiten am Hof in Peking', in *Europa und die Kaiser von China 1240–1861*. Frankfurt am Main: Insel Verlag, 107–13 (exhibition catalogue).

1985d 'Map-making and geography traced in the David M. Stewart Collection', in *The Discovery of the World: Maps of the Earth and the Cosmos from the David M. Stewart Collection*. Montreal: David M Stewart Museum, 17–28.

1985e *Raleigh and Roanoke, the First English Colony in America, 1584–1590: The British Library Exhibition Hosted by the North Carolina Museum of History*. Raleigh: North Carolina Department of Cultural Resources (exhibition catalogue).

1985f 'The role of the painter in Renaissance marine cartography', in *Imago et Mensura Mundi: Atta del IX Congresso Internazionale di Storia delta Cartografia*, vol. 2. Rome: Istituto della Encyclopedia Italiana, 515–23.

1986a (with P. Barber) 'Anonymous, 1665–1670, Island of Manhodos/The Towne of New York, and John Miller, 1695, New Yorke', in English *Mapping of America: An Informal Selection of Printed and Manuscript Maps Produced during the Formative Years of the English Map Trade*. Mercator Society Publication 1. New York: The Mercator Society, New York Public Library.

1986b (with P. Barber) 'Cartographic materials relating to Germany in the British Library', in D. Paisey (ed.), *German Studies, British Resources*: Papers *Presented at a Colloquium at the British Library 25–27 Sept 1985*. British Library Occasional Papers 8. London: British Library, 62–5.

1986c (with D. Middleton) 'The mapping and exploration of Africa and Joseph Banks', in I. Sternberg and P. M. Larby (eds), *African Studies*: Papers *Presented at a Colloquium at the British Library 7–9 Jan 1985. British* Library Occasional Papers 6. London: British Library, 188–9.

1986d 'Material on nautical cartography in the British Library, 1550–1650', *Revista da* Universidade de Coimbra 32, 187–97.

1986e '"So geographers in Afric-maps"', *Map Collector* 36, 30–4.

1986f 'South Asian maps in the British Library Reference Division', in A. Gaur (ed.), *South Asian Studies*: Papers *Presented at a Colloquium at the British Library 24–26 April 1985. British* Library Occasional Papers 7. London: British Library, 49–55.

1986g '"Things hidden from other men": The Portuguese voyages of discovery', *History Today* 36, 27–33.

1987a (ed. with A. H. Robinson) *Cartographical Innovations: An International Handbook of Mapping Terms to 1900*. Tring: Map Collector Publications in association with the International Cartographic Association.

1987b 'Memorial to Edmund Halley (1656–1742)', *Map Collector* 38, 44.

1987c (with Anna Maria Crinò) 'New Researches on the Molyneux Globes', *Der Globusfreund* 35–7, 11–18 (German summary 19–20).

1988a 'Did the Portuguese discover Australia?' *History Today* 38, 30–5.

1988b 'Java la Grande: The enigma of the Dieppe maps', in G. Williams and A. Frost (eds), *Terra Australis to Australia*. Melbourne: Oxford University Press, 38–81.

1988c 'Maps, charts and diplomacy in the Glorious Revolution', in *1688*: *The Seaborne Alliance and Diplomatic Revolution*: *Proceedings of an International Symposium held at the National Maritime Museum*, Greenwich, *5–6 Oct 1988*. Greenwich: Trustees of the National Maritime Museum, 151–64.

1989a 'Further comments on the "lost harbour"', *Map Collector* 49, 34–5.

1989b 'Historical cartography', in D. W. Rhind and D. R. F. Taylor (eds), *Cartography Past, Present and Future*. Amsterdam, ICA and Elsevier Applied Science Publications, 111–22.

1989c '19de-eeuwse thematische kartografie', *Caert-Thresoor* 8(2), 46–52.

1989d 'Opera Mundi: Emery Molyneux, Jodocus Hondius and the first English globes', in T. C. van Uchelen, K. van der Horst and G. Schilder (eds), *Theatrum Orbis Librorum. Liber Amicorum Presented to Nico Israel on the Occasion of His Seventieth Birthday*. Utrecht: HES Publishers, 94–104.

1989e 'A Portuguese discovery? The enigma of the Dieppe Maps', in J. H. and A. Frost (eds), *Studies from Terra Australis to Australia*. Canberra: Australian Academy of the Humanities *Occasional* Papers 6, 46–55.

1989f with G. Schilder 'Speed military maps discovered', *Map Collector* 48, 22–6.

1990a 'Charts of the sixteenth-century schools of seamanship', *Atti del V Convegno Internazionale di Studi Colombiani: Navi e navigazione nei secoli XV e XVI, Geneva, 26–28 ottobre 1987.*

1990b 'Maps of the American War of Independence', *IMCoS Journal* 42, 7–15.

1990c 'The Vinland map: Fake, forgery or *jeu d'esprit*?', *Map Collector* 53, 2–6.

1991a (with P. S. Grove) 'Discovery of the Rawlinson copperplate maps of the Americas and their related prints', *Map Collector* 56, 12–21.

1991b 'Earliest visions of the Pacific: Conference keynote address', in *Proceedings of the Great Ocean Conferences: The North Pacific*. Portland: North Pacific Studies Center, 1–15.

1991c 'Globes', in *The Christopher Columbus Encyclopedia*, vol. 1. New York: Simon & Schuster, 297–301.

1991d 'Northwest Passage', in *The Christopher Columbus Encyclopedia*, vol. 2. New York: Simon & Schuster, 518–22.

1991e 'Intercourse with the peaceful muses', in S. Roach (ed.), *Across the Narrow Seas: Studies in the History and Bibliography of Britain and the Low Countries: Presented to Anna E. C. Simoni*. London: British Library, 31–54.

1991f 'The Vinland map: Genuine or fake?', *Bulletin du bibliophile* 1, 76–83.

1992a 'George Anson's voyage, 1740–44', *IMCoS Journal* 50, 6–12.

1992b 'Cartographic knowledge of the world in 1492', *Mariners Mirror* 78, 407–18.

1992c 'The challenge that is an Australian map', in *The Globe Unfolding Australia* (Journal of the Australian Map Circle, Special Issue) 37, 1–11.

1992d 'The Portuguese debate. Java-La-Grande and Australia', *The Globe Unfolding Australia* (Journal of the Australian Map Circle, Special Issue), 12–13.

1992e 'Is the Paris map the long-sought chart of Christopher Columbus?', *Map Collector* 58, 20–1.

1992f 'Vignettes of the cartography of Cyprus', *First Agnes Michaelides Memorial Lecture*. Nicosia: Cultural Foundation of the Bank of Cyprus.

1992g 'What Columbus knew', *History Today* 42, 17–23.

1992h 'World libraries and museums', in A. Rogers, H. Viles and A. Goudie (eds), *The Students' Companion to Geography*. Oxford: Blackwell Publications, 287–92.

1993a 'Cartography and cosmography in the age of discovery: The measure of the Earth from Ptolemy to Columbus', in A. Lafuente, A. Elena and M. L. Ortega (eds), *Mundializatiân de la tienda y cultura national, Attas del Congreso Internacional 'Ciencia, descubrimiento y mundo colonial*. Madrid: Ediciones Doce Calles, 75–80.

1993b 'Discovery at Holkham Hall', *Colonial Williamsburg* 15(3), 49–55.

1993c (with M. A. Lowenthal) 'The missing print in the Rawlinson–Holkham series on the Americas', *Map Collector* 63, 10–11.

1993d 'What the eye can't see ...', in P. Barber and C. Board (eds), *Tales from the Map Room*. London: BBC Books, 18–19.

1993e 'Fantasy and misunderstanding', in P. Barber and C. Board (eds), *Tales from the Map Room*. London: BBC Books, 36–7.

1993f 'Informing the people back home', in P. Barber and C. Board (eds), *Tales from the Map Room*. London: BBC Books, 120–1.

1993g *William P. Cumming Map Collection at Davidson College*. Davidson: Davidson College.

1994a (with M. H. Edney) 'Cartography', in I. Grattan-Guiness (ed.), *Companion Encyclopedia of the History and Philosophy of the Mathematical Sciences*, vol. 2. London and New York: John Hopkins University Press, 1101–14.

1994b 'From South Sea to Pacific Ocean. Changing conceptions of the Pacific Ocean in the late 18th century', *Navigation: Newsletter of the Australian* Institute of Navigation 2(1), 1–6.

1994c (co-editor with A. McConnell) *Historian's Guide to Early Maps of the British Isles*. London: Royal Historical Society.

1994d 'Navigators and mathematical practitioners in Samuel Pepys's day', *Journal of Navigation* 47, 1–19.

1994e 'Rotz, Jean', in C. S. Nicholls (ed.), *Dictionary of National Biography: Missing Persons*. Oxford and New York: Oxford University Press, 572.

1995 'The great publication societies', in G. Warkentin (ed.), *Critical Issues Editing Exploration Texts*. Toronto: University of Toronto Press, 108–24.

1996 'Sixteenth century maritime manuscript atlases for special presentation', in J. A. Wolter and R. E. Grim (eds), *Images of the World: Essays on the History of the Atlas*. Washington: Library of Congress.

1999 *British Globes up to 1850: A Provisional Inventory*, ed. R. Dunn (based on the draft catalogue entries and notes compiled by Helen Wallis). London: British Library.

Selected biographical sources and works on Helen Wallis with contextual works

1995a 16 May BL HWA: F. G. Emmison, then County Archivist of Essex, to the author (pers. comm.) From another letter to HW in 1961, from a person signing as 'Stoddart', I deduce that Helen was to have been married on or about Saturday 8 July 1961. A photograph of her fiancé is in the BL HW archive.

1995b Commemorative essays in the obituary of Eila Muriel Joyce Campbell (1915–1994), *Imago Mundi* 47(1), 7–12.

1999 http://www.mccroneinstitute.org/v/66/McCrone-The-Vinland-Map (consulted 16 October 2018).

Anon. [T. Campbell], (1995), 'Obituary', *The Times*, 17 February.

Bailes, H. (1995) contribution in S. Tyacke and T. Campbell (eds), *The Globe My World: Tributes to Helen Wallis*. London: Robin Alston (privately printed).

Bailes, H. (2000), *Once a Paulina*. London: James & James, 134–5.

Campbell, T. (1995), 'Helen Margaret Wallis OBE (1924–1995): Obituary and select bibliography', *Imago Mundi* 47, 185–92 (also includes record of many obituaries not cited separately here).

Campbell, T. (1997), [contribution], 7–17 in *St Hugh's Memorial Seminar 12 April*, circulated to friends and others. Cited in text as 1997 *St Hugh's* (privately printed).

Campbell, T. (2004), 'Wallis, Helen Margaret', in H. C. G. Matthew and B. H. Harrison (eds), *Oxford Dictionary of National Biography*. Oxford: Oxford University Press; consulted 8 October 2018 (also includes record of many obituaries not cited separately here).

Campbell, T. and Tyacke, S. (1994), 'Dr Helen Wallis (1924–95) former Map librarian of the British Library reminiscing about her career in maps with Tony Campbell and Sarah Tyacke on 5 August 1994', https://youtu.be/EbMMIgEGeCQ, cited in the text as 'HW video 1994'.

Edney, M. (2014), 'Academic cartography, internal map history and the critical study of mapping processes', *Imago Mundi* 66(Supp.), 83–105.

Edney, M. (2018), OML Curatorial records series 2, Box 19 Collections: Helen Wallis book collection and other acquisitions (1995–2004).

Hall, B. (1995), [contribution] S. Tyacke and T. Campbell (eds), *The Globe My World: Tributes to Helen Wallis*. London: Robin Alston (privately printed), 14.

Hall, B. (1997a) [contribution], *St Hugh's Memorial Seminar 12 April* (privately printed), 15.

Hall, B. (1997b) [contribution], *St Hugh's Memorial Seminar 12 April* (privately printed), 28–30.

Harley J. B. (1987), 'Innovation social context and the history of cartography', *Cartographica* 24(4), 59–68.

Harvey P. D. A. (2006), 'The Vinland map. R A. Skelton and Josef Fischer', *Imago Mundi* 58(1), 95–9.

Howse, D. and Gueritz, E. (1995), 'Obituary', *Mariner's Mirror* 81(2), 131–3.

Mead, W. R. (1995a), 'Obituary', *Geographical Journal* 161(2), 240–1.

Mead, W. R. (1995b), 'Obituary', *Independent*, 14 February.

Richardson, W. (2016), *The Portuguese Discovery of Australia: Fact or Fiction*. Canberra: National Library of Australia.

Saunders, A. (1997), *St Hugh's Memorial Seminar 12 April* (privately printed), 14.

Scott, V. and Campbell, T. (1995), *Map Collector* 70, 43.

Tyacke, S., ed. (1994), *A Celebration of the Life and Work of Eila M. J. Campbell.* London: Robin Alston (privately printed).

Tyacke, S. (1995), *Guardian*, 24 February.

Tyacke, S. and Campbell T., eds (1995), *The Globe My World: Tributes to Helen Wallis.* London, Robin Alston (privately printed).

Webb, M. (1995), [contribution] in S. Tyacke and T. Campbell (eds), *The Globe My World: Tributes to Helen Wallis.* London: Robin Alston (privately printed), 18–19.

Woodward, D. (1997), [contribution], *St Hugh's Memorial Seminar 12 April* (privately printed), 35–43.

Chronology

1924	Born in Barnet Hertfordshire 17 Aug
1934	Attended St Paul's School, London
1945	Matriculated as geography student, St Hugh's College, University of Oxford
1949	BA honours Geography
1951	Appointed Assistant Keeper (Class 2) BM Map Room
1967	Appointed Deputy Keeper, as Superintendent of the Map Room
1971–88	Chairman, Society for Nautical Research
1972–4	President, British Cartographic Society
1973	Became BL Map Librarian when BM Library departments became part of the BL
1977–81	Chairman, Geography and Map Libraries section of International Federation of Library Associations
1979	Senior Research Fellow, William Andrews Clarke Library, University of California, Los Angeles
1982	Honorary Vice-President, RGS
1985	Officer of the Order of the British Empire
1986	Retired from the BL
1986–94	Council Member, Foundation for Science and Technology
1986–95	Member, Archival Research Task Force examining the voyages (1574, 1577, 1578) of Martin Frobisher
1986–95	President, International Map Collectors' Society
1993–4	First Mitchell Library (State Library of New South Wales) scholar in residence
1995	Died, 7 Feb

1995	Funeral, St John's Wood Church, 10 Feb
1995	Commemorative meeting, RGS, 9 May
1995	Posthumously awarded the Victoria medal of the RGS
1995	Memorial service, St George's Church, Bloomsbury
1997	Memorial seminar, St Hugh's College, Oxford, 12 April

5

Isabel Margarida André, 1956–2017

Diogo Gaspar Silva

Figure 5.1 Isabel André, Jubilee of Professor Orlando Ribeiro, 1981 Faculty of Letters, University of Lisbon. Photography by Ilídio do Amaral.

Education, life and work

Isabel Margarida de Almeida André was born on 30 July 1956 in the city of Lisbon in Portugal. The only daughter of António Dias André, clerk at the Portuguese National Navigation Company, and Ana Margarida Duarte de Almeida André, clerk at Metallized Metallurgical Societies, Isabel had a happy childhood in Queluz in the greater Lisbon region. She was very close to her extended family and had a particularly close relationship with her father, whose death was a major blow to her. She remained close to her paternal grandmother and often played at her grandfather's pharmacy in Queluz.

André was a studious and successful pupil from elementary school through to her studies at the old high school of Queluz. Growing up in the Second Portuguese Republic, a right-wing, authoritarian regime, the adolescent André developed a special interest in social issues. Her own profound determination and sense of freedom were perhaps influenced by her relationship with her father, a member of the then illegal Portuguese Communist Party. André had her first child, a daughter, in 1973 aged 17 and as a mother did not join volunteers of the student civic service, set up in 1975 by the new democratic Portuguese government to help reconstruct the country after its long period of dictatorship.

André entered university just after dictatorship in Portugal gave way to democracy. She wavered between studying geography or economics: her partner and the father of her daughter, Pedro Macedo, was an economics student and possibly influenced André's choices at this stage. In the end, however, she entered the Department of Geography of the Faculty of Letters of the University of Lisbon (FLUL–UL) in the academic year 1975–1976. Her chosen department offered a very traditional education, modelled closely on the French school with its emphasis on regional geography. André chafed at this regional emphasis, which remained in place during her time in the department, but one of its teachers and her later colleague, Jorge Gaspar, was already adopting modern perspectives and this interested André, as it did her great friend Patrícia Rêgo, later the author of an emotional memoir of André in the Portuguese geography journal *Finisterra* (Rêgo 2017). New geographical paradigms were emerging at the same time as the political situation was rapidly becoming more liberal, and in the new political climate, Portuguese university courses became much more flexible and André was able to take optional courses that matched her pre-existing interests: political geography, with Jorge Gaspar, and another in economics. In 1980 André finished her five-year geography degree, graduating with a classification of 17 out of 20 (A grade). The same year she began her teaching career as assistant trainee in the Department of Geography at the invitation of Jorge Gaspar.

André had already begun in 1978 as a young researcher in the Center for Geographical Studies of the University of Lisbon (CEG–UL), founded by Orlando Ribeiro, called Studies for Regional and Urban Planning, coordinated by Jorge Gaspar. Her early research was inspired by ideas of the 'New geography' coming into Portugal from outside in the new period of openness. In addition to new methods, such as those of the quantitative revolution, there was a focus on applied geography, offering practical solutions to concrete problems, and a shift in scale of analysis from the local, associated with the French school, to the global. Inspired by these developments, which fitted her deeply held political views, André began her career and scientific contribution to electoral studies in the 1980s.

Personal geography

André formed not only professional but also personal links to several places with which her connections have been long-lasting. In the first phase, from the 1980s, André's relations expanded to Spain, namely to Madrid, where she had several friends with whom she also worked on issues of planning and evaluation of public policies.

In the early 1990s, André was one of the founders of Geoideia, a planning company founded by a group of seven young geographers and a civil engineer who included, among others, Nuno Costa, João Ferrão and Jorge Malheiros. The context for its foundation was the new emphasis on territorial planning and the evaluation of public policies after the entry of Portugal into the European Economic Community in 1986. The company structure allowed more freedom of operation, creativity and collaboration than was possible in the still rather hierarchical university department. Its studies in demography, socio-territorial issues, location strategies, questions of gender and territory, and territorial planning helped the development of territorial management at the municipal level in the democratic regime.

At the end of the 1980s and in the 1990s André consolidated her links with Spain, particularly Barcelona, where she developed contacts with Maria Dolors García-Ramon, who inspired her interest in gender issues in the 1990s. Barcelona also played a key role in her work in the twenty-first century on the problems of arts and creativity. In Brussels André worked with João Ferrão on the analysis and evaluation of public policies.

With the creation of the Foundation for Science and Technology in Portugal in 1997, André began to focus on projects at the national level, starting a phase of projects more focused on academia. Some of them, such as KATARSIS or SOCIAL POLIS dating from the beginning of the twenty-first century, took André to France, Austria and, above all, Canada, in particular through the contribution of Frank Moulaert and Juan-Luis Klein. Her focus was on public space and creativity as a form of urban regeneration and innovation. The circus, in particular, was an enduring personal and professional interest: as a child she loved the circus and continued to visit it with her daughter and grandchildren, and it was also the subject of her academic work (André and Reis 2009).

Having earlier spent extensive vacations in Galicia with Spanish and French friends whom she had known in the 1980s and in the 1990s, André turned to Portugal for periods of recreation. Within Portugal, it was the Azores and particularly the island of Faial, in whose volcanic landscapes André enjoyed summer holidays until about 15 years before her death, that best integrate André's personal and professional geographies. Montemor-o-Novo and later Évora, both cities in Alentejo, Portugal, also feature in André's personal geography with studies of socio-territorial innovation and creativity and with the presence of her long-time friend at the University of Évora, Patrícia Rêgo.

Scientific ideas and geographical thought

Isabel André's research fields were closely integrated with questions and places, such as Madrid, Barcelona, Paris, Montreal, Horta, Montemor-o-Novo and, more recently, Évora, which were important to her personally. Moreover, a retrospect of her scientific contribution shows paradigmatic shifts, but reveals that new ventures always flowed from her previous experience and long-standing interests (Figure 5.2). In addition to being a teacher and a researcher, she was active in planning, consulting and evaluation of public policies, via her company Geoideia.

Epistemological transition: breaks and continuities

a. Political and electoral geographies

From 1978 André began her research in the group of Studies for Regional and Urban Planning at FLUL–UL. The transition from fascism to democracy, with all its associated transformations, stimulated her research in the 1970s and

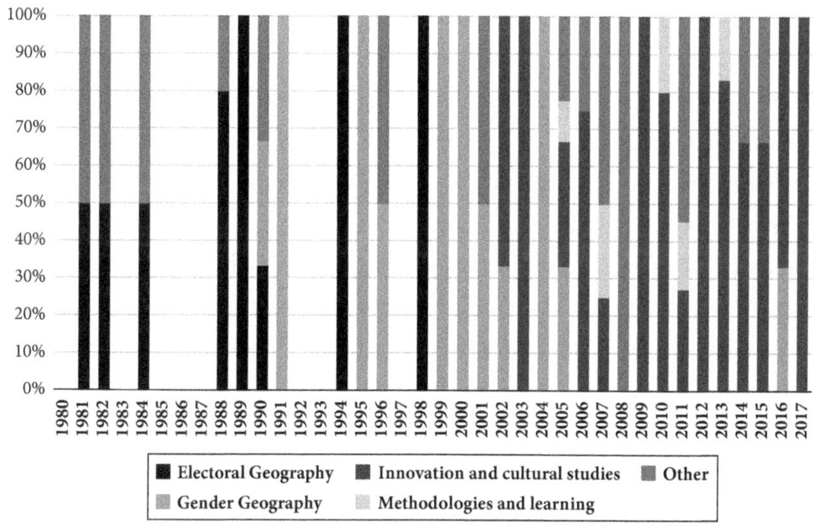

Figure 5.2 Isabel André's scientific life: breaks and continuities (1980–2017).

1980s. In the early 1980s André participated in the first studies of electoral geography in Portugal with Jorge Gaspar and Fernando Honório. Her report on the electoral results of 1980 (André and Honório 1981) was followed by studies published at the invitation of the Social Research Institute Damião de Góis, in Lisbon, on municipal and legislative elections respectively (André, Gaspar and Honório 1982a, 1984a). These studies attempted a 'deeper and more comprehensive analysis' (André, Gaspar and Honório 1984a, 3) of the behaviour of the Portuguese electorate, namely, analysis of voting, the configuration of political systems and the transfer of votes between elections. In local elections she concludes that voting patterns are explained by a combination of the personal relationship between the voter and the representative, and approval or disapproval of the policies followed at the central level (Silva and André 1994a). This conclusion is also mirrored in her 1984 master's dissertation 'Local power: Elections, elected and municipal management: contribution to the geography of local power in Portugal' ('O poder local: Eleições, eleitos e gestão municipal: Contributo para a geografia do poder local em Portugal'). In this dissertation, supervised by Jorge Gaspar, André analyses politicians and voters in the district of Santarém and whether local power could be a counterweight to the central state. She notes that 'the electorate seems to tend to value the previous rulers, a situation that can reflect the importance of the relations that are established between the municipalities and the citizens and especially between them and the local economic agents' (André 1984b, 85), so personal factors trump ideological ones. She found this phenomenon again ten years later in 'Local power' ('Le pouvoir local') (Silva and André 1994a). Throughout the dissertation, André emphasizes financial decentralization at national and local levels, and the fragility of formal democracy, also evident at the national level, given the many professional, technical staff who are elected and the possibility that this could undermine voters' relationship with institutions. André also concludes that municipal investments are more concentrated in collective infrastructure and equipment than are those of the central state (André 1984b; Silva and André 1994a). She also analysed local finances, building on her two earlier articles on 'Regional financial flows' ('Fluxos financeiros regionais') (André 1982b) and 'Geography of banking institutions in Portugal, 1974–1988' ('Geografia das instituições bancárias em Portugal, 1974–1988') (André and Malheiros 1990a).

Further pioneering studies sought to elucidate trends in electoral geography, for example, her work with Carlos Patrício (1984c), 'The different images of geography' ('As diferentes imagens da geografia'), a work also characterized by her great sense of humour. Perhaps her definitive analysis in the field came with the publication of 'Catholicism in Portugal: Ecclesiastical organization and religious practices' ('O catolicismo em Portugal: Organização eclesiástica e práticas religiosas') (André and Patrício 1988a) or 'Urban electoral behaviour in legislative elections' ('O comportamento eleitoral urbano nas eleições legislativas') (André 1988b).

André explained electoral behaviour by the greater or lesser influence of capitalism. For example, in the case of Lisbon, Setúbal and Santarém, 'the communist vote strongly opposes not only abstention but above all voting for other parties' (André 1988b, 64), a trend that can be explained by the 'deepening and extension of the capitalist social relations that characterize this region' (Andre 1988b, 65). Another aspect of her analysis was the classification of social groups proposed by João Ferrão, which is evident in her master's dissertation in which she differentiated the electorate theoretically and empirically according to their conditions of work, the rural–urban dichotomy, the ageing of the population, and religious practices (André 1988b). She found religious practices to be more intense 'in regions where pre-capitalist economic forms or cultural references are associated with them' (André and Patrício 1988a, 247), emphasizing the importance of the family. Conversely, in urban–industrial areas, Catholicism is less dominant and support for the Communist Party higher. Regions with the largest number of catholic missionaries or marriages show above-average support for the right-wing CDS–PP and PSD – a conclusion on which all empirical quantitative studies in the 1980s seem to converge and which was highlighted in a more global and integrated study entitled *Portugal: electoral geography: 1975–1987* (*Portugal: Geografia eleitoral: 1975–1987*) (Gaspar and André 1989).

André's interest in political and especially electoral geography endured until the end of her life, and coloured her teaching as well as her research from the beginning of the 1980s at FLUL–UL and later at the Institute of Geography and Spatial Planning (IGOT–UL). Electoral issues were widely discussed in her political geography course, and her students appreciated her enthusiasm for the topic.

b. Gender geography and the feminist perspective

André was one of the first to introduce gender studies to Portuguese geography, being part of the new critical geography that had emerged over the previous three decades, and gender themes dominated her life in the 1990s. André explains in her 1990 article in *Finisterra* the importance of incorporating social issues into geographic analysis, and suggesting that Marxist approaches often ignored the spatial component. Feminist geography thus emerged with a focus on 'social differences and inequalities between men and women and their consequences in the organization and use of space' (André 1990b, 336), challenging the validity of using average values as was prevalent in human geographical studies at the time.

Writing as a mother of nearly twenty years' experience, André states that it is 'women who continue to provide domestic work, family assistance and education for children' (André 1990b, 337), and she points to the dichotomies that explain the social (in)visibility of women: public/private, production/reproduction, work/leisure and formal work/informal work. She integrates into her analysis processes that define the social relations between the genders, as well as the functioning of the family institution, arguing that patriarchal and capitalist systems of stratification must be integrated to understand the relationship between gender and class.

André went on to identify the factors that characterize gender inequality in access to the labour market, an analysis that extends throughout the 1990s (André 1991). Pointing to the strength of Catholicism in Portugal during the dictatorship, André identifies the factors that led to the emancipation of women in society in general and in the labour force in particular. These included industrialization and the consequent rural exodus, which promoted change in family models; involvement of men in the colonial wars after the end of fascism which led to women's entering the workforce in large numbers; and particularly, improvements in the education of girls and women. However, there were setbacks: in the aftermath of decolonization, some 500,000 immigrants returned to Portugal and in consequence the female unemployment rate soared in the 1980s, while the prevalence of neoliberalism led to more flexible work contracts and an associated rise in precariousness. The growth of the private sector in Portugal demonstrated the difficulties faced by working women: the flexibility demanded in the labour market was incompatible with women's status as providers of domestic tasks and reproduction (André 1991).

André identifies these factors in many of her published works and in her doctoral dissertation 'The false neutral in human geography: Gender and patriarchal relationship in employment and domestic work' ('O falso neutro em geografia humana: género e relação patriarcal no emprego e no trabalho doméstico') (1994b), supervised by João Ferrão, with whom she shared various interests (Figure 5.3). This work received the 1995 Women's Research Award Carolina Michaelis de Vasconcelos, promoted by the Commission for Equality and Women's Rights, an entity with which André collaborated over the years. More recently her work emphasized the great weight of domestic work that limits women's geographical and socio-professional options and leaves men better able to compete in the flexible and neoliberal labour market (André and Brito-Henriques 1995a; André 1996).

This research was pioneering in the Portuguese context where gender differences had generally been ignored and a wholly unrealistic gender neutrality and equality assumed. Inspired by Jorge Gaspar's classes in political geography 'where the neutral had no place' (André 1994b, 7), André emphasizes the growing interest in social inequality, political attitudes and cultural practices

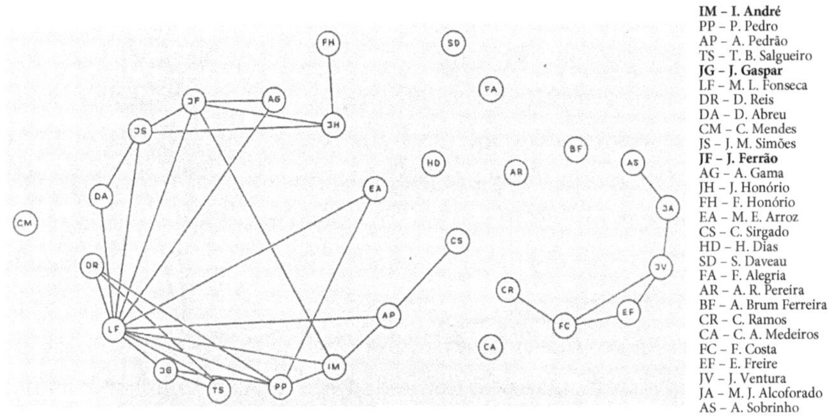

Figure 5.3 Positioning CEG–UL members from the cited authors.
Source: Book of Homage to Orlando Ribeiro, CEG–UL, vol. 1, 1988, 52.

(André and Brito-Henriques 1995a), which rejected gender neutrality and identified mobility as a factor in inequality. This new geographical approach provoked criticism and disquiet in some geographers and André confronted their resistance (1990b, 1994b) to defend a socially engaged geography, where gender was not a neutral condition.

André's argument and concerns are framed in a context of postmodernity, in which class struggle had given way to movements based on identity politics, although it also recognizes the weakness of feminist movements in Portugal, and the 'non-neutral' character of individuals is affirmed in a global context. These factors justify 'the expansion of the struggle for the elimination of all forms of discrimination' (1994b, 11). Growing individualism, and concerns about equality and difference challenged traditional family structures and the ubiquity of marriage as the precursor to household formation – things also favoured by late capitalist productive relationships.

As André detailed in her doctoral dissertation, the flexibility of post-Fordist capitalism affected gender relations, 'not only by attenuating some differences in the job market, but also because flexibility in work processes imposes new flexibilities on social reproduction which encompasses aspects as varied as the modes of consumption and the division of domestic tasks between the spouses' (1994b, 73). She based her conclusions on her empirical identification of nine professional profiles. She first identifies two large sets of professional profiles with respective low or advanced levels of education; infrequent or frequent on the job training, short-term or permanent employment contracts and little or clear career progression. Then, introducing other variables such as current profession, employing institution, unionization or reasons for unemployment in the last year, André identifies nine professional profiles. Women have a significant presence in four profiles: the unemployed; low status and precarious employment (profiles 1 and 2), employment in the public sector (profile 6), and in activities related to school education (profiles 8 especially teachers and 9 especially nurses). Then, fourteen different domestic work profiles were drawn, considering the sharing of domestic work between family members and the family model of domestic work organization. For example, tasks related to cleaning and laundry are mainly carried out by the women themselves, except in the upper profiles (profiles 11 to 14), where they are done by a maid. Tasks related to electrical repairs are performed by the woman in only 5 per cent of the clusters. Weekly shopping is a joint task, since it is done by car with the woman depending on the husband's driving licence. Her dissertation finally relates professional profiles to a typology of domestic work. André (1994b) concludes, for example, that women with low qualifications, precarious employment contracts and low professional progression (profiles 1 to 4) in most cases also do non-mechanized domestic tasks, do not engage others to do the household tasks and take a disproportionate share of such tasks. By contrast, women with more skilled jobs (profiles 5 and 9) tend to engage others to perform many domestic tasks, sharing those that remain more equally with their male partners. Where a strict Marxist perspective would argue that patriarchal relationships were the consequence of the capitalist relationships, which determine all domains of society, André argues that capitalism alone could not explain gender inequalities, and that capitalism and patriarchy are autonomous social processes (André 1994b).

André's 'At the center or the periphery?' (1996) confirms the hypotheses she had advanced at the beginning of the decade. Comparing the working conditions of men and women in the European Union of twelve nations, she argues that Portuguese women tend to have a central role in family life, because, unlike in some European countries, the income of women, even those with small children, is necessary for the survival of the family. This was particularly the case during Portugal's colonial war. She also identifies the overrepresentation of women in the textile industry and in social and personal services – functions traditionally performed by women within the domestic sphere. Nevertheless, in the 1980s and 1990s the difference in employment of men and women was remarkable. Gender differentiation has since been blurred as education opened to women and enabled them to improve their lot – something which reduced the patriarchal configuration of society. Labour practices thus go a long way to explaining the differences in countries' gender relations.

At the beginning of the 1990s, through João Ferrão, André met Ana Nunes de Almeida, a researcher at the Institute of Social Sciences of University of Lisbon (ICS–UL). They collaborated on several work projects, resulting from their shared interest in children's issues, violence in the family and school failure (André, Almeida and Almeida 1999a, 2000, 2001).

André considered that gender inequality was primarily explained by inequalities within the family and their articulation with other organizations of society, such as labour and business, since capitalist development led to the separation of the spheres of production and reproduction, and subordinated the second to the first. Consequently, the division of labour brought about the modern nuclear family model. André participated in several studies examining these ideas including *Recent Patterns of Fertility in Portugal* (*Os padrões recentes da fecundidade em Portugal*) (André et al. 1995b); 'Shadows and marks: The mistreatment of children in families' ('Sombras e marcas: Os maus tratos às crianças nas famílias') (André, Almeida and Almeida 1999a); *Socio-Familial Contexts of Child Abuse* (*Os contextos sociofamiliares dos maus tratos às crianças*) (André, Almeida and Almeida 2000) and 'Sons and daughters: A different relationship with school' ('Filhos e filhas: Uma diferente relação com a escola') (André, Almeida and Cunha 2005a). These focused on the need to understand the dynamics of different family models in the asymmetries of fertility behaviours and the school performance of boys and girls.

André's deep knowledge of connections among family structure, sexuality, conjugality and fertility was essential to understanding abuse in certain family contexts. She argued for example that the weakening of emotional and family ties led to increased aggression against children in families marked by physical violence in the most deprived fringes of the population and to affective and psychological neglect in more prosperous families. For these reasons, André joined several projects related to changes in nuptiality and sexuality and, consequently, in family models, namely 'Gender inequality and European regions' (1995–1997), coordinated by Simon Duncan; 'Families in contemporary Portugal: Structures, dynamics and solidarities' (André and Feio 1999b; André, Almeida and Almeida 1999a, 2000) (1997–2001), coordinated by Karin Wall at the University of Lisbon; 'Men, fertility and contraception: Reproductive health life courses' (André et al. 2002; André et al. 2004a; Almeida and André 2004b) (2002–2004), coordinated by Ana Nunes de Almeida at the University of Lisbon; and, finally, 'Gender and mobility space–time inequality' (2015–2016) at CEG–UL, coordinated by Margarida Queirós.

c. *Innovation and cultural studies*

André's interest in cohesion and socio-territorial innovation began under the influence of Jorge Malheiros's doctoral thesis 'Migratory archipelagos: transnationalism and innovation ('Arquipélagos migratórios: transnacionalismo e inovação') (2001) and rapidly led to publications (André 2002). She led and coordinated several national and European projects focusing on urban issues. Her focus was initially on social and spatial cohesion through the analysis of the networks of places, through the project 'LINKS: Social capital and innovation, networks promoting local development' (2004–2008), based in CEG–UL, and, from 2005, she began to explore how culture and the arts can induce urban regeneration and socio-territorial innovation. Among her main projects, in addition to those in which she participated as a team member, were 'KATARSIS: Growing inequality and social innovation: Alternative knowledge and practice in overcoming social exclusion in Europe' (2006–2009), coordinated by Frank Moulaert and Jean Hillier, and 'SOCIAL POLIS: Social platform on cities and social cohesion' (2008–2010), again coordinated by Frank Moulaert. Both projects had Isabel André as a coordination partner in Lisbon. Projects dedicated to investigating how culture and the arts promote the development and cohesion of places were 'RUcaS: Real utopias and socially creative spaces' (2011–2014), and 'ÁGORA: Encounters between the city and arts: Exploring new urbanities' (2016–2019), both coordinated by André and based in CEG–UL.

André's interest was initially in the entrepreneurial aspect of cultural renewal and she argued that innovation should be considered as an investment and not as a cost to be avoided. However, such studies rapidly gave way to a focus on issues of cohesion and innovation at the scale of the whole city. André's attention was captured by the liveliness and rapid transformation of cities. After the April 1974 revolution, public space returned to its place at the centre of civic life, especially in times of political or financial crisis. André argued that it was in times of crisis that innovations, such as social ones, emerged.

In 2005 André supervised Eduardo Medeiros's dissertation 'Territorial cohesion in the NUTS III of the continental Portugal border: the INTERREG–A initiative and regional development: The case of NUTS III of Central Alentejo' ('A coesão territorial nas NUTS III de fronteira de Portugal Continental: a iniciativa INTERREG-A e o Desenvolvimento Regional: O caso da NUT III do Alentejo Central'). (NUTS or Nomenclature of Territorial Units for Statistics is a European Union system for referencing the subdivisions of countries. NUT III, in the Portuguese case, represents a group of municipalities. INTERREG-A is a European Commission project that supports cooperation between NUTS III regions on national borders). The following year André developed a study on frontier regions and projects of cooperation and socio-territorial cohesion between Portugal and Spain through the analysis of the INTERREG programme (André and Moreira 2006a).

It was, however, in 2006 that André developed her own approach to the issues of social innovation, significantly influenced by Frank Moulaert and Juan-Luis Klein. Together with the economist Alexandre Abreu, she revisited the concept of innovation, focusing on a social perspective, and analysing processes, agents and phenomena, including spatial ones, related to social innovation (André and Abreu 2006b). André assumes that social innovation, unlike technological innovation, is not commercial but rather collective, and has the capacity to generate and transform social relations. As she had done throughout her career, she tackled issues of inequality and vulnerable groups. In this way she argued that social innovation resulted from the process of social inclusion and can empower weaker agents, where gender is included. According to André and Abreu (2006b), same-sex marriage is an example of social innovation in the field of ethics (André 2013a). They also give the example of people who resort to micro-credit, who want to meet unmet human needs through the market, contributing to the social inclusion of these individuals. For that reason, these examples show that social innovation involves the empowerment of the beneficiaries, triggering a greater or lesser change in the power relations in which they were involved (André and Abreu 2006b).

In this last intellectual phase André responded to Richard Florida's vision of a creative and inclusive city, where tolerance, technology and talent come together in utopian harmony. Social innovation, including cultural and artistic innovation, overcomes difficulties and short-term risks, and is predominantly non-commercial and led by an informal social movement. For this reason, 'The conceptualization of social innovation cannot avoid the subject of power relations and their transformation' (André and Abreu 2006b, 130). Social innovation, according to André, can occur only in socially creative milieux, which can undergo transformations without losing their identity. In this way, 'A milieu cannot be creative–innovative if it penalizes the possible failures of a risky initiative, that is, if it is too rigid, hierarchical and normative' (André and Abreu 2006b, 132). This terminology is discussed in the doctoral dissertation of André's student Ana Estevens (2014) 'Geographic expressions of urban conflict: The cases of Raval (Barcelona) and Mouraria (Lisbon)' ('As expressões geográficas do conflito urbano. Os casos do Raval (Barcelona) e da Mouraria (Lisboa)').

Having discussed concepts and ways of quantifying social innovation, André dedicated the years after 2014 to investigating the arts as a mechanism of social innovation and as a catalyst for the regeneration of run-down urban areas. In this phase Barcelona re-entered André's personal geography, through the study of creative social strategies in the case of the Walden–7 (an apartment building designed by Ricardo Bofill in the late 1960s which combined counter hegemonic architectural and social experiences and whose name was inspired by a utopian community from the science fiction novel *Walden Two*, itself an allusion to Thoreau's *Walden* (1854)) (André and Rouselle 2010a), but above all Montreal and Paris dominate the last years of her life.

André often travelled to Canada from where she got inspiration for her aggregation thesis 'The circus in the city: Cirque du Soleil on a round trip to Quebec, Montreal' ('O circo na cidade: O Cirque du Soleil numa viagem de ida e volta ao Québec, Montreal') (2008). (André was until 2008 an associate teacher. She wrote her aggregation thesis to gain eligibility for higher teaching positions; having successfully defended the thesis, in 2009, she was promoted to associate teacher with aggregation just one level below 'Full-Teacher' rank). Her thesis focused on active teaching–learning issues. The allusion to Cirque du Soleil, the Montreal-based '*nouveau cirque*' (new circus) company, reappeared in André's article 'The circus comes to the city! Opportunities for socio-territorial innovation' ('O circo chegou à cidade! Oportunidades de inovação sócio-territorial') (2009), which analyses the role of Chapitô in Lisbon (a state funded school for performing arts and crafts, theatre company, a restaurant, a live music bar, a daycare centre, and circus school for socially vulnerable young people) as a case study of social innovation. In this context 'the circus is an artistic expression with unique potential for participating in the construction of socially creative cities. It does not have a history related to the elites' (André and Reis 2009, 81). Cirque du Soleil is again analysed in the case of Montreal's inner-city Saint-Michel neighbourhood where it originated, and she describes it as a project with strong social commitment in terms of social inclusion, but still a megaproject of urban regeneration in a peripheral and problematic part of Montreal.

The focus on the circus recalls André's childhood experience and coincides with the birth of her grandchildren at the beginning of the century, in both of which periods the circus played a large part in her life. This is another example of epistemological continuities, since André consistently studied what she lived, but the circus was also attractive in being a socially creative milieu, fostering plurality, diversity and conviviality among generations. It also exemplified tolerance, mobilization and identity, being, in the case of Chapitô, a socio-territorial innovation 'both through the improvement of the urban public space and through the social inclusion of children and young people at risk' (André and Reis 2009, 79).

Public space is a recurring topic and features in the master's dissertation of André's student Aquilino Machado, 'Public spaces in Lisbon: The Portuguese World Exhibition and Expo98' ('Espaços públicos em Lisboa: A Exposição do Mundo Português e a Expo98') (2003). André and Machado discuss how the public space can deepen citizenship through new civic practices anchored in creativity and the arts, and thus constitute a focus of social innovation (Machado and André 2012). Analysing the Marais in Paris, the city André liked to explore on foot, the authors identify the public space as a projector of civil society, where artistic and cultural performances foster links between different communities, promote a more diversified tolerant and civically committed local community, and establish the dialogue that makes the Marais a socially creative neighbourhood.

André coordinated several projects aimed at understanding how the arts and culture regenerate and develop places. She was strongly inspired by

Frank Moulaert, a great friend with whom she worked on several projects on innovation and cultural studies.

Having participated actively in projects with similar objectives, namely DivesT (2000–2004), LINKS (2004–2007), KATARSIS (2006–2009) and SOCIAL POLIS (2008–2010), André coordinated RUcaS between 2011 and 2014. Using case studies, including Montemor-o-Novo and other new urban areas, the project investigated the relationship between innovation and creativity and how the arts can contribute to development and inclusiveness. A major publication from this project was André's 'Inspiration and emotions: Culture and arts engendering new urban places' (André 2017a). André intended to produce a set of public policy guidelines for urban rehabilitation and regeneration projects. In addition to being a consultant in several projects, such as 'Cross-border cooperation and local development in rural space of Alentejo/Extremadura' ('Cooperação transfronteiriça e desenvolvimento local no espaço rural do Alentejo/Extremadura') (2004–2006), or 'Cities of tomorrow: Challenges, visions and ways forward' ('Cidades de Amanhã: desafios, visões e perspetivas'), in 2010, André evaluated public policy programmes through which highlighted the issues of knowledge, innovation (technological and social) and creativity from 2009 onwards.

The RUcaS project fostered links with an international network of research on cultural and innovation studies with poles in Belgium, Quebec and Neuchatel in Switzerland – all places important to André personally in this last phase of her career – and reinforced links with colleagues like Frank Moulaert or Juan-Luis Klein who were consultants on the project.

This project gave rise to an innovative publication titled *Atlas of Real Utopias: Creativity, Culture and the Arts* (*Atlas das utopias reais: Criatividade, cultura e artes*) (2016) where the case studies, namely Mouraria, Vale da Amoreira, LxFactory, Montemor-o-Novo and Montreal, illustrate the promotion of creative, competitive and inclusive places, highlighting that 'creativity emerges associated with the arts, creative industries, the mobilization of creatives for urban regeneration, urban development and the right to the city' (André, Estevens and Gabriel 2016, 8). Emotional geography and creativity foster cohesion and social inclusion to produce real utopias in the neoliberal city. It is typical of André that this atlas has a vision for the future, as illustrated by her chapter on the evolution of artistic education which seeks to identify who are the creatives of the city (André, Estevens and Gabriel 2016).

The last project that André coordinated, which continued with Ana Estevens as coordinator, is entitled ÁGORA. Its title alludes to the open space and discussion of the Athenian polis. Through the arts, the right to the city becomes a form of valorization or even protest. With case studies including the Almirante Reis-Colina de Santana Avenue, Chiado, Alvalade, Mouraria, Marvila-Xabregas, all in Lisbon city, Barreiro (Lisbon Metropolitan Area), Rua Miguel Bombarda in Porto, and places in Madrid, Athens and Dublin, the project seeks to understand how, in times of crisis, the neoliberalization of the state and austerity, public urban spaces can be sites of contest and resistance to neoliberal transformations (Machado and André 2012). This analysis has its roots in the revolutionary periods after 25 April 1974, when public space was the locus of resistance – something that made a lasting impression on André.

In addition to coordinating these two projects, from 2011 André was coordinator of the research group 'ZOE: Dynamics and urban and regional policies' in CEG–UL, which investigated smart, sustainable and inclusive solutions for urban and regional development as well as citizenship in a global and urbanized world. At the same time, ZOE analyses the contemporary transformations and how financial crisis and the associated austerity policies affected urban and regional dynamics. This last aspect is directly related to the objectives triggered by the RUcaS, ÁGORA and other projects in which André participated. It sought to understand how transformations and resistances take place in the city through the arts, artists and cultural agents. These projects tried to understand how arts, artists and cultural agents have been shaped by financial crisis and austerity policies and how those elements generated the transformation of the contemporary cities by producing new urbanities and new cultural scenes as a form of resistance.

Active teaching–learning methodologies: a contribution to an inclusive education

Reflective by nature and determined by conviction, André from the 1990s was critical of accepted pedagogies as part of her long-standing concern for knowledge transmission. She found the democratization of teaching incompatible with traditional teaching methods and thought that educational institutions should reinvent themselves in response to new political, social and above all technological realities. Frequent discussions with Herculano Cachinho, with whom she shared an office in the Interdisciplinary Complex and in FLUL–UL during the 1990s, prompted an interest in methodological issues and active teaching–learning and her work *Research Methodologies in Human Geography* (*Metodologias de investigação em geografia humana*) (2005b) described 'a set of personal convictions about how the teaching of Geography in the University should be conducted' and affirmed that 'a Geography course (undergraduate or postgraduate) should be a place where the knowledge and the pleasure of discovery are privileged through active and creative learning processes' (André 2005b, 5).

An interest in methodological issues cuts across all André's works as far back as the 1990s when, in collaboration with her last life partner, Fernando João Moreira, she developed the first programme on theory and methodologies in geography in the 1994–1995 academic year at the 'Research Seminar in Geography' ('Seminário de Investigação em Geografia') taken by undergraduates in their final year as they undertake their final graduation project. She embraced new methodologies appropriate to specific problems, from quantitative methods to urban living labs and even augmented reality. In the study commissioned by the Centre for Judicial Studies entitled *Families and Child Abuse in Portugal* (*Famílias e maus tratos às crianças em Portugal*) (André, Almeida and Almeida 2001), for example, the construction of a database enabled the use of factor analysis and cluster analysis, which in turn defined categories for different family models. This interest and knowledge are further portrayed in her doctoral thesis and in the categories of family organizations (André 1994b, 2005c), and in studies on human development and social cohesion (André 2005d). Her interest in methodological and active teaching–learning issues led to André's participation in a conference in Newcastle on problem-based learning. In the graduate seminar 'Culture and socio-territorial innovation', André started implementing problem-based learning, an experience that led to her aggregation thesis about Cirque du Soleil's return to Montreal. The seminar and the problem-based method that André incorporated, called for students to 'identify the factors – strategies, policies and practices – that make the territories socially creative' (André 2008, 3). Urban living labs or action research appeared later in her teaching (André et al. 2017b).

André was also a member of the Scientific Committee of the Observatory of Student Pathways of the University of Lisbon, presenting her concerns to new publics and especially with the new forms of teaching and learning that the democratization of access to higher education and online methods made possible. Her collaboration with ICS–UL resulted in the publication of several studies on school drop-out and failure, including *UL Caloiros: A year later* (*Caloiros da UL: Um ano depois*) (André et al. 2010b), *Students at the Entrance of UL* (*Os estudantes à entrada da UL*) (André et al. 2011) and the concern with these issues in the IGOT–UL, with the title 'School failure and drop-out in IGOT–UL: Contributions to inform action' ('Insucesso e abandono escolar no IGOT–UL: Contributos para informar a ação') (André and Cachinho 2013b). André also supervised master's dissertations on pedagogical issues and inclusive teaching (Mendes 2007; Pereira 2008). André was a great advocate of the independence of the IGOT–UL, something that was formalized to her great pleasure in 2008.

As both teacher and researcher, André was concerned with inclusion. She always valued the young researchers with whom she worked and with whom she developed close collaborative relationships. Among them, Ana Estevens, André Carmo and Leandro Gabriel and the economist Alexandre Abreu stand out.

Isabel André: the personification of freedom, critical thinking and inclusion

The life and scientific contribution of Isabel Margarida André are marked by freedom, critical spirit and inclusion. Her yearning for freedom is the result of the period and environment in which she grew up, where political discussion and the deconstruction of political discourses were unceasing, especially in her family circle. Electoral geography, particularly Portuguese electoral behaviour immediately after the April Revolution, conducted with Jorge Gaspar and Fernando Honório, was her main focus in the 1980s. Her master's dissertation in 1984 contributed to the understanding of the characteristics of electoral behaviour in Santarém and highlighted the difficulties with local finances that stemmed from Portugal's sociopolitical transformation since 1974 and underlay the voting patterns which she identified.

As young mother, alive to social and political issues in a family context, André turned to the power relations of capitalism and patriarchy throughout the 1990s with her pioneering work on gender studies in Portugal, which led to prominence on the international stage. The first phase took her to Spain, mainly Madrid, with friends who joined her projects to evaluate spatial planning in the company Geoideia, and later to Barcelona. Influenced by pioneering feminist Maria Dolors García-Ramon who inspired André's doctoral dissertation, she emphasized the need for a spatial analysis of gender. Her doctoral dissertation 'The false neutral in human geography' ('O falso neutro em geografia humana'), with its thesis that neutrality is impossible, stemmed from her 'academic formation, but also from personal convictions' (1994, 7). In the same decade André participated in studies with the ICS–UL focusing on fecundity and domestic violence and, a little later, school failure and drop-out, with Ana Nunes de Almeida. André argued that family organization and its structure were essential for understanding these phenomena of exclusion.

Also in the 1990s and influenced by Herculano Cachinho, André embraced active teaching and learning, aimed at the new audience that wider access to higher education had created. School failure and new ways of producing knowledge, based on critical and reflexive thinking, preoccupied her as she sought to harness the benefits of new forms of communication and access to knowledge.

From 2005 André focused on urban and cultural issues and extended her contacts to Canada via her link with Juan-Luis Klein. By consolidating new interests with her personal experiences, through study of the circus, she made the city her laboratory to show how the arts and culture can foster socially creative spaces, where innovation and creativity combine to disrupt and contest current political paradigms. Her last two projects, RUcaS and ÁGORA, show public space as the locus for freedom, inclusion and innovation.

André died on 3 April 2017, aged 60, after a prolonged illness, at her home, in Estoril, Portugal. André left behind her mother, her current husband, Fernando João Moreira, her daughter, Ana Margarida André, and five grandchildren, two girls and three boys. After a service on 4 April 2017 in the Church of Queluz, her funeral ceremony took place on 5 April 2017 in the Barcarena cemetery, where she was cremated.

Figure 5.4 Isabel André, a restless geographer (Isabel André, uma geógrafa inquieta), ZOE Spring Meeting, May 2018.
Source: IGOT–UL

Since then, memorial events have taken place in her memory. On 18 May 2018 ZOE spring meeting was held in IGOT–UL with a commemorative conference (Figure 5.4). In the presence of her family, friends and colleagues, the conference 'Isabel André: A restless geographer' ('Isabel André: uma geógrafa inquieta') featured a roundtable discussion on the main lines of her engagement and research: electoral geography by Jorge Gaspar; gender studies by Ana Nunes de Almeida; local and regional development by João Ferrão; culture, arts and social innovation by Frank Moulaert; and education by Herculano Cachinho. This conference was very emotional, especially during the showing of a short film about André created by French director Pierre Primetens. At the end, Ana Estevens and André Carmo, both young researchers who worked with André from the beginning, together with Fernando João Moreira, launched a commemorative volume with the same title as the conference. An exhibition with personal photographs and information was held at the IGOT–UL until the end of May 2018. Finally, on 11 October 2018, celebrations to mark 10 years of the IGOT–UL included a panel discussion dedicated to Isabel André, entitled 'Gender, inequalities and territory' (Figure 5.5). Among the attenders were Maria Dolors García-Ramón, one of André's great inspirations on this

Figure 5.5 Gender, inequalities and territory panel discussion, 10th Anniversary of IGOT, October 2018.
Source: IGOT–UL

topic, and a representative of the Commission for Citizenship and Gender Equality (Comissão para a Cidadania e Igualdade de Género), an organization, with which André collaborated over the years.

For these reasons, her legacy in the social sciences in general and in geography in particular is her inspiration to others to imagine what might be possible, to struggle for freedom, critical thinking and inclusion, and to seek utopia, something epitomized in her last published work: 'Inspiration and emotions: Culture and arts Engendering new urban places' (André 2017a). Frank Moulaert (2017, 5) wrote that 'Isabel was a great, good and warm person, a creative scholar, a wonderful teacher and tutor'. Her scholarly rigour secures her a place in geography, and her openness makes her much missed.

Acknowledgements

I thank Ana Nunes de Almeida and Patrícia Rêgo for their interviews that were essential for the writing of this essay, which also benefitted from the collaboration and comments of Ana Estevens, Aquilino Machado, Herculano Cachinho and Jorge Malheiros.

Geographical context of Isabel André

Almeida, A. N., André, I. and Almeida, H. N. (2002), 'Os maus tratos às crianças na família', *Acta Médica Portuguesa* 15, 257–67.

Estevens, A. (2014), 'As expressões geográficas do conflito urbano: Os casos do Raval (Barcelona) e da Mouraria (Lisboa)', unpublished doctoral dissertation, Universidade de Lisboa.

Machado, A. (2003), 'Espaços públicos em Lisboa: A Exposição do Mundo Português e a Expo98', unpublished master's dissertation, Universidade de Lisboa.

Malheiros, J. (2001), 'Arquipélagos migratórios: Transnacionalismo e inovação', unpublished doctoral dissertation, Universidade de Lisboa.

Medeiros, E. (2005), 'A coesão territorial nas NUTS III de fronteira de Portugal continental: A iniciativa INTERREG-A e o desenvolvimento regional: O caso da NUT III do Alentejo Central', unpublished master's dissertation, Universidade de Lisboa.

Mendes, S. (2007), 'Educação e desenvolvimento: as consequências do abandono escolar precoce na inserção na vida ativa: Estudo de caso sobre o ensino básico no concelho de Beja', unpublished master's dissertation in Development, Local Diversity and Global Challenges - Analysis and Management, ISCTE-Instituto Universitário de Lisboa.

Moulaert, F. (2017), 'Introduction', in P. van den Broeck, C. Parra and A. Mehmood (eds), *Thoughts for a Better World: Social Innovation as Political Transformation*. Leuven: European Spatial Development Planning Network, 5.

Pereira, J. (2008), 'Inclusão dos alunos das comunidades ciganas nas escolas portuguesas', unpublished master's dissertation, Universidade de Lisboa.

Rêgo, P. (2017), 'A geógrafa Isabel André, notas de testemunho de um percurso inovador, brilhante e livre', *Finisterra* 52/105, 143–5.

Selected publications by Isabel André

1981 with Honório, F., 'Os resultados eleitorais de 1980: Legislativas e presidenciais', *Finisterra* 16/31, 199–211.

1982a with Gaspar, J. and Honório, F. *As eleições para as câmaras municipais: Estudo de geografia eleitoral*. Lisbon: Instituto de Pesquisa Social Damião de Góis.

1982b 'Fluxos financeiros regionais', *Finisterra* 17/34, 199–211.

1984a with Gaspar, J. and Honório, F. *As eleições para a Assembleia da República 1979–1983: Estudo de geografia eleitoral*. Lisbon: Instituto de Pesquisa Social Damião de Góis.

1984b 'O poder local: Eleições, eleitos e gestão municipal: Contributo para a geografia do poder local em Portugal', unpublished master's dissertation, Universidade de Lisboa, Lisbon.

1984c with Patrício, C., 'As diferentes imagens da geografia', *Finisterra* 17/34, 391–4.

1988a with Patrício, C., 'O catolicismo em Portugal: Organização eclesiástica e práticas religiosas: Análise regional', *Finisterra* 23/46, 225–49.

1988b 'O comportamento eleitoral urbano nas eleições legislativas', *Povos e Culturas* 3, 227–43.

1989 with Gaspar, J., 'Portugal: Geografia eleitoral: 1975–1987', in M. Coelho (ed.), *Portugal: O sistema política e constitucional 1974–1987*. Lisbon: Instituto de Ciências Sociais, 257–77.

1990a with Malheiros, J., 'Geografia das instituições bancárias em Portugal, 1974–1988', *Finisterra* 25/49, 227–43.

1990b 'O género em geografia: Introdução de um novo tema', *Finisterra* 15/50, 331–48.

1991 'Women's employment in Portugal', *Iberian Studies* 20(1 and 2), 28–41.

1994a with Nunes Silva, C., 'Le pouvoir local', *Peuples Mediterranées* 66, 91–102.

1994b 'O falso neutro em geografia humana: Género e relação patriarcal no emprego e no trabalho doméstico', unpublished doctoral dissertation, Universidade de Lisboa.

1995a with Brito-Henriques, E., Práticas de lazer, território e género. VI Colóquio Ibérico da Geografia. Atas., vol. 2, Porto: Publicações da Universidade do Porto, 665–72.

1995b with Almeida, A. N., Ferreira, C. and Ferrão, F., *Os padrões recentes da fecundidade em Portugal*. Lisbon: Comissão para a Igualdade e para os Direitos das Mulheres.

1996 'At the centre or the periphery? Women in the Portuguese labour market', in M. García-Ramon and J. Monk (eds), *Women of the European Union: The Politics of Work and Daily Life*. London: Routledge, 138–55.

1999a with Almeida, A. N. and Almeida, H. N., 'Sombras e marcas: os maus tratos às crianças nas famílias', *Análise Social* 150, 91–122.

1999b with Feio, P., 'Development and equality between women and men in the Portuguese labour market', *South European Society and Politics* 4(2), 54–71.

2000 with Almeida, A. N. and Almeida, H. N., *Os contextos sociofamiliares dos maus tratos às crianças*. Atas do III Congresso Português de Sociologia. Oeiras: Celta Editora.

2001 with Almeida, A. N. and Almeida, H. N., *Famílias e maus tratos às crianças em Portugal, Assembleia da República*. Centro de Estudos Judiciários, Lisbon: Divisão de Edições.

2002 'Coesão espacial: A inclusão dos lugares nas redes do desenvolvimento', in H. Cachinho, I. André, I. Medeiros, J. Reis, J. Malheiros, M. H. Esteves, R. J. Marques, S. Claudino and T. Barata Salgueiro (eds), *Olhares sobre o território e espacialidade*. Estudos de Geografia Humana e Regional 45, Lisbon: Centro de Estudos Geográficos, 43–56.

2004a with Almeida, A. N., Vilar, D. and Lalanda, P., *Fecundidade e contraceção: Percursos de saúde reprodutiva das mulheres portuguesas*. Lisbon: Imprensa de Ciências Sociais.

2004b with Almeida, A. N., 'O país e a família através das crianças: Ensaio exploratório', *Revista de Estudos Demográficos* 35, 5–35.

2005a with Almeida, A. N. and Cunha, V. 'Filhos e filhas: Uma diferença relação com a escola', in K. Wall (ed.), *Famílias em Portugal: Percursos, interações, redes sociais*. Lisbon: Imprensa de Ciências Sociais, 517–52.

2005b *Metodologias de investigação em geografia humana*. Estudos de Geografia Humana e Regional 47, Lisbon: Centro de Estudos Geográficos.

2005c 'Família e género', in C. A. Medeiros, T. Barata Salgueiro and J. Ferrão (eds), *Geografia de Portugal: Sociedade, paisagens e cidades*, vol. 2. Lisbon: Círculo de Leitores, 72–86.

2005d 'Desenvolvimento humano e coesão social', in C. A. Medeiros, T. Barata Salgueiro and J. Ferrão (eds), *Geografia de Portugal: Sociedade, paisagens e cidades*, vol. 2. Lisbon: Círculo de Leitores, 141–50.

2006a with Moreira, F. J., 'A progressiva aproximação das duas margens da raia', in P. Rêgo (ed.), *Aproximar as margens: Cooperação transfronteiriça e desenvolvimento local no espaço rural do Alentejo e da Extremadura*. Évora: Fundação Luís de Molina, 96–102.

2006b with Abreu, A., 'Dimensões e espaços da inovação social', *Finisterra* 41(81), 121–41.

2008 'O circo na cidade: O Cirque du Soleil numa viagem da ida e volta ao Québec, Montrea', unpublished aggregation thesis, Universidade de Lisboa.

2009 with Reis, J., 'O circo chegou à cidade! Oportunidades de inovação socioterritorial', *Finisterra* 44(88), 79–94.

2012 with Machado, A., 'O espaço público do Marais-Paris', *Finisterra* 47(94), 119–36.

2010a with Rouselle, M., 'Estratégias sociais criativas em Barcelona: O caso do Walden-7', *Finisterra* 45(90), 71–90.

2010b with Almeida, A. N., Vieira, M. M. and Alves, N., *Caloiros da UL: Um ano depois*. Observatório dos Percursos dos Estudantes (OPEST), Lisbon: Reitoria da Universidade de Lisboa.

2011 with Almeida, A. N., Vieira, M. M., Alves, N. and Oliveira, V. *Os estudantes à entrada da UL*. Observatório dos Percursos dos Estudantes (OPEST). Lisbon: Reitoria da Universidade de Lisboa.

2013a 'Gender and social innovation: The role of EU policies', in F. Moulaert, D. MacCallum, A. Mehmood and A. Hamdouch (eds), *International Handbook on Social Innovation: Collective Action, Social Learning and Transdisciplinary Research*. Aldershot and Brookfield: Edward Elgar, 412–26.

2013b with Cachinho, H., 'Insucesso e abandono escolar no IGOT–UL: Contributos para informar a ação', in A. N. Almeida (ed.), *Sucesso, insucesso e abandono na Universidade de Lisboa: Cenários e percursos*. Lisbon: EDUCA, 93–116

2016 with Estevens, A. and Gabriel, L. (co-ordinators), *Atlas das utopias reais: Criatividade, cultura e artes*. Lisbon: CEG/Le Monde Diplomatique.

2017a 'Inspiration and emotions: Culture and arts engendering new urban places', in P. V. den Broeck, C. Parra and A. Mehmood (eds), *Thoughts for a Better World: Social Innovation as Political Transformation*. Leuven: European Spatial Development Planning Network, 94–101.

2017b with Estevens, A. and Gabriel, L., 'Urban living labs: Cultural scenes, the arts, and artists as levers of new urbanities', *Canadian Journal of Regional Science* 40(1), 43–52.

[n.d.] Encontros entre a cidade e as artes: explorando novas urbanidades [Official webpage of the ÁGORA project, coordinated by André], https://agoraprojecto.wordpress.com/ (accessed 17 October–13 December 2017).

[n.d.] Utopias reais em espaços socialmente criativos [Official webpage of the RucaS project, coordinated by André], https://agoraprojecto.wordpress.com/ (accessed 17 October–13 December 2017).

Chronology

1956	Born in Lisbon, on 30 July
1966	Elementary school: Externato Almeida Garrett
1968	High school: Liceu Passos Manuel (Massamá extension)
1969	High school: Liceu Nacional de Queluz
1972	High school: Liceu Rainha Dona Leonor
1974	Matriculated at the University of Lisbon, Faculty of Letters
1975	Joins the geography course
1977	Member of Centre for Geographical Studies of University of Lisbon, being part of EPRU research group (Urban and Regional Planning Studies)

1978	Geography Department Monitor (Faculty of Letters, University of Lisbon)
1980	Graduates in Geography Trainee Assistant in the Faculty of Letters, University of Lisbon
1981	Member of the Assembly of Representatives in the Faculty of Letters, University of Lisbon
1984	Graduated with master's degree in human geography Assistant in the Faculty of Letters, University of Lisbon
1990	Founder member of Geoideia – Territorial organization studies, Lda. Development of planning, consultancy and public policy evaluation, such as INTERREG, LEADER and EQUAL
1993	PhD in human geography, University of Lisbon
1994	Assistant Teacher in Faculty of Letters, University of Lisbon
1995	Carolina Michaelis de Vasconcelos Women Research Award for PhD thesis Team member of the project 'Internationalization, economic restructuration and the production of new marginalities', coordinated by T. Barata-Salgueiro, University of Lisbon Network member of the project 'Gender inequality and European regions', coordinated by Simon Duncan, University of Bradford
1996	Network member 'Knowledge intensive services and innovation', coordinated by Peter Wood, University College London
1997	Joined Regional Geography research group at Centre for Geographical Studies Team member of the project 'Families in contemporary Portugal: Structures, dynamics and solidarities', coordinated by Karin Wall, University of Lisbon
1998	Team member of the project 'Employment and employability in retailing', coordinated by T. Barata-Salgueiro, University of Lisbon Team member of the project 'Reproductive health of women in Portugal', coordinated by Ana Nunes de Almeida, University of Lisbon
2002	Team member of the project 'Men, fertility and contraception: Reproductive health life courses', coordinated by Ana Nunes de Almeida, University of Lisbon
2004	Associate Teacher in Faculty of Letters, University of Lisbon Coordinator of the project 'LINKS: Social capital and innovation, networks promoting local development'
2006	Team member of the project 'KATARSIS: growing inequality and social innovation: alternative knowledge and practice in overcoming social exclusion in Europe', coordinated by Frank Moulaert and Jean Hillier, University of Newcastle
2007	Team member of the project 'Success and unsuccess in the University of Lisbon: Sceneries and paths', coordinated by Ana Nunes de Almeida, University of Lisbon

2008	Important contribution to the foundation of the Institute of Geography and Spatial Planning Team member of the project 'SOCIAL POLIS: social platform on cities and social cohesion', coordinated by Frank Moulaert, University of Newcastle and KU Leuven
2009	Associate Teacher with Aggregation at Institute of Geography and Spatial Planning, University of Lisbon
2011	With Mário Vale, University of Lisbon, coordinator of the project 'RUCAS: Real utopias and social creative spaces' Coordinator of the research group NEST: Strategies and territorial policies nucleus
2015	Coordinator of the project ÁGORA: Encounters between the city and arts: Exploring new urbanities Team member of the project 'GENMOB: Gender Mobility: Time–space inequality' Coordinator of the research group 'ZOE: Urban and regional change and policies'
2017	Died in Estoril, 3 April

Index

This index comprises:

1. A general index, including person names, country names, institutions, organizations, and some thematic entries
2. A cumulative index of biobibliographies included in this and previous volumes of Geographers.

General Index

Cumulative Index